Engineering and Sustainable Development

Environmental Science, Engineering and Technology

nova
science publishers
www.novapublishers.com

Engineering and Sustainable Development

Intelligent Decision Support System for IoT-Enabling Technologies: Opportunities, Challenges and Applications
Subrata Sahana, PhD (Editor)
Anil Kumar Sagar, PhD (Editor)
Sanjoy Das, PhD (Editor)
Vishal Jain, PhD (Editor)
2024. ISBN: 979-8-89113-249-8 (Hardcover)
2024. ISBN: 979-8-89113-333-4 (eBook)

More information about this series can be found at
https://novapublishers.com/product-category/series/engineering-and-sustainable-development/

Environmental Science, Engineering and Technology

Conserving, Halting and Reinstating Degraded Natural Resources During the UN Decade on Ecosystem Restoration
Rajani Srivastava, M.Sc., PhD (Editor)
Pankaj Kumar, PhD (Editor)
2023. ISBN: 979-8-89113-067-8 (Hardcover)
2023. ISBN: 979-8-89113-140-8 (eBook)

More information about this series can be found at
https://novapublishers.com/product-category/series/environmental-science-engineering-and-technology/

Hussain H. Al-Kayiem
Hasanain A. Abdul Wahhab
Raed A. Jessam
Editors

Advancements in Sustainability Systems

nova
science publishers
www.novapublishers.com

DOI: https://doi.org/10.52305/PURZ5664

NOTICE TO THE READER

Library of Congress Cataloging-in-Publication Data

Names: Al-Kayiem, H. H. (Hussain Hammod), editor. | Wahhab, Hasanain A. Abdul, editor. | Jessam, Raed A., editor.
Title: Advancements in sustainability systems / editors, Hussain H. Al-Kayiem, PhD, professor, Hilla University College, Babylon, Iraq, Hassanain A. Abdul Wahhab, PhD, University of Technology, Baghdad, Iraq, Training and Workshop Centre, Raed A. Jessam, PhD, University of Technology, Baghdad, Iraq, Electromechanical Engineering Department.
Description: New York : Nova Science Publishers, [2024] | Series: Engineering and sustainable development | Includes bibliographical references and index. |
Identifiers: LCCN 2024009149 (print) | LCCN 2024009150 (ebook) | ISBN 9798891135710 (hardcover) | ISBN 9798891136465 (adobe pdf)
Subjects: LCSH: Sustainable engineering. | Sustainable development.
Classification: LCC TA163 .A375 2024 (print) | LCC TA163 (ebook) | DDC 628--dc23/eng/20240314
LC record available at https://lccn.loc.gov/2024009149
LC ebook record available at https://lccn.loc.gov/2024009150

Published by Nova Science Publishers, Inc. † New York

Contents

Preface

Solutions to many global drawbacks are imbedded in the adoption of the seventeen identified sustainable development goals (SDGs). The current book compiles materials split into 10 chapters that provide data, information, historical aspects, findings, prospectives and identification of related matters and topics in some of SDGs. In particular, the techniques and systems for human health and safety, the urban development systems and approaches, and energy systems for water and power generation have been the main three topics considered in the book.

Urban quality is a strategic issue for smart futuristic cities as it represents a basic goal for sustainable city development and the formation of a sustainable urban space. Development of cities towards improving environmental quality and self-perception of human value, improving, and developing urban quality is explained. It discusses the space-urban quality synchronization, the multi-dimensional construct, and the major factor that affects the quality of urban life and the creation of attractive places that give the greatest value to the communities and achieve luxury. In addition, the impact of generative algorithms in different stages of the urban design approach is explained by qualitative measurements in both collecting and analyzing data from the literature to study the shifting in approaches down from the traditional to the contemporary methods in urban design. A detailed unique case of the application of generative design in Kartal Sub-Center and Kartal-Pendik Waterfront Urban Transformation in Istanbul metropolitan area on the Asian continent of Turkey is described with analysis. In addition, the importance of functional land use planning, how it changes over time, and the reasons for these changes to achieve an acceptable standard of living for residents, also known as the concept of livability, is very well discussed. Analyses of the related materials showed that the goals of functional land planning could be achieved through optimal functional distribution that is aligned with sustainability aspects by reducing noise and the negative externalities of cars and relying on walking. It has been concluded that this sustainability aspect could be achieved through the functional proximity of services to each other within a specified range.

5G technology has a numerous role on cleaner environment and sustainability. The modern high-tech communication by 5G technology is also taken part in this book through proposed solution to the noise generated in the fiber optic communication systems. Filter bank multicarrier (FBMC) is considered one of the methods that improve the impairment in orthogonal frequency division multiplexing (OFDM), such as Cyclic Prefix (CP), which takes place in bandwidth without uses. The parallel analysis filter bank with modified blind phase search estimation (PAFB-MBPS) is proposed and analyzed to improve the spectral efficiency and increase the quality factor due to using full bandwidth.

Human health is one of the widely attracted the attention of developers, policy makers, designers, researchers, and investors. The critical and significance of ventilation systems in preventing the spread of infectious diseases within multifunctional buildings is directly related to human health and wellbeing. Part of the book examines various ventilation strategies, emphasizing their impact on indoor air quality and occupant safety. Indoor settings by interdisciplinary interaction among architects, engineers, and healthcare specialists, linking public health concerns with broader sustainability aims. Optimizing ventilation systems not only minimizes disease transmission, but also adds to a healthier interior ecosystem, promoting health and environmental sustainability, as well as energy saving. The critical and long-term implications of engineering control measures on indoor air quality, thermal control, and environmental impact sustain healthier living and working environments. This perspective aligns with the overarching goal of creating healthier indoor spaces while also promoting environmentally responsible practices for a more sustainable future. The chapter also discusses the control measures to reduce or eliminate the spread of viruses in closed environments. A critical issue on health and security issues related to personal well-being is cranioplasty, as it is one of the essential aims of sustainable development of health, justice, and prosperity. The objective of cranioplasty is to protect the brain after cranial surgery and ensure sustainable functioning of the brain. Furthermore, it alleviates psychological difficulties, while improving social performance.

No escape from solar systems as a major approach towards clean and affordable sustainable energy and water production. The presented materials represent state of the art on the solar PV application. The book considers compiled the radiative cooling technology as state of the art in the PV solar panels cooling by increasing the heat emissivity or spectral selectivity of the incident light. It also highlighted spectrum selectivity technology, which works by reflecting undesirable light while letting in the needed light. The most renowned literature on radiative cooling and solar spectrum management techniques was looked over to understand more about the advantages and potential of this type of PV cooling. An interesting approach for sustainable clean water production, near nonclean water resources, by harnessing solar energy is explored and compared. It is without reservation could be said that solar-water generation is an effective solution to address global clean water scarcity. It evaluates a modified double-slope solar still compared to conventional solar stills, showcasing significant improvements in distillate output and energy efficiency. This part of the book discusses modifications like finned basins, heat storage materials, solar collector, wick materials, and external enhancements like reflectors. This comprehensive resource offers valuable insights for sustainable water production, ensuring reliable access to clean water for communities in need. However, in areas with no water resources, the book presents and discusses the use of solar energy and thermoelectric technology for sustainable water production in remote and off-grid areas. Solar energy offers an economical and eco-friendly solution, eliminating the need for grid connection to operate water purification/production systems. Integrating solar PV and thermoelectric generators into water production systems enhances the atmospheric water generation (AWG) and makes it feasible, even in arid regions. AWG systems with thermoelectric generators can be adapted for various capacities, addressing water scarcity. Development of this approach meets the UNDP SDG for sustainable clean water production.

The book, could be said, is a reference point that addresses important solutions, data, techniques, and discussions on various sustainability systems development.

Chapter 1

An Outlook on Sustainable Development of Urban Quality and Place

Maryam M. Al-Aarajy[1,*]
and Nada A. Hasan[1]
[1] University of Technology- Iraq, Baghdad, Iraq

Abstract

The urban environment has lost, at present, many of its features and operational capabilities and is suffering from ever-increasing and aggravating problems, causing economic, social, demographic, urban, and cultural problems, which made controlling these problems and formulating policies and solutions a difficult matter that disturbs the planning and executive authorities in the city, thus negatively affecting the well-being of the human. Therefore, understanding the requirements for achieving urban quality and the interrelated factors that influence it helps us work together to improve the quality of urban city environments. Urban quality can be understood as a specific relationship linked to the place that represents the basic structure of the city, as the factors, characteristics, and dimensions of the place are essential to understanding the concept of urban quality and the demand for high-quality urban environments with good urban design, which is seen as a tool to reshape the built environment to accommodate the new urban conditions. Therefore, the current research aims to explore and clarify the relationship of urban quality to place by providing a textual analysis of the concept of urban quality in general and its relationship to place in particular. By discussing the main research axes and analyzing a group of studies, the researchers concluded that the concept of urban quality is

---

* Corresponding Author's Email: maryam.mohammedah@gmail.com.

In: Advancements in Sustainability Systems
Editors: Hussain H. Al-Kayiem, Hasanain A. Abdul Wahhab and Raed A, Jessam
ISBN: 979-8-89113-571-0

complex, with various features, relative, subjective, and dynamic, and changes with the introduction of new models. These features could be achieved when the environments are rich in values, and this is what creates beneficial places socially, economically, and environmentally, with distinct characteristics that bring luxury and comfort to its users. The spatial factors (the integration of tangible and intangible physical factors, non-physical, cognitive, and perceptual factors), spatial characteristics (social, behavioral, and physical), and spatial dimensions (formal, social, cultural, functional, visual, and sensory), as indicators of assessing urban quality that affect the quality of places and urban life and achieve satisfaction with the place.

Keywords: urban quality, place, spatial factors, spatial characteristics, spatial dimensions, urban sustainable development

1. Introduction

Global cities are developing towards improving quality and self-perception of human value (Ye et al., 2021), and in particular, improving and developing urban quality that reflects the strength of the city and represents the quality of its urban development, as cities are among the important determinants of future sustainability and human well-being (Gomaa & Fouad, 2022; Yousif & Al-Hinkawi, 2022). The concept of urban quality is linked to the concept of sustainable urban development, as urban quality with sustainable development requires an interactive effort between humans and nature in the urban environment, and this requires the integration of the economic, social, urban, and psychological fields that affect humans and living organisms directly or indirectly, to achieve a healthy and livable environment. There is an overlap between the concept of urban quality and the concept of sustainability. Sustainability focuses on the future dimension, the balance of the relationship between humans and the environment, and the right of future generations to enjoy the environment. In contrast, urban quality focuses on the current spatial and temporal dimensions (Turkoglu, 2014).

Studies indicated that urban quality represents an imperative that bears the essence of Place design and planning and supports all processes of space production and shaping urban Places (Gomaa & Fouad, 2022). It involves understanding, relating, analyzing, and synthesizing space and spatial attributes from diverse perspectives through a relational epistemology of space's physical and social forms and uses (Khan et al., 2014). The problem of the research was represented by the suffering of the urban environment from exacerbating and ever-increasing problems at the level of place and the lack of a comprehensive theoretical framework that explains the details of urban quality and its relationship to place and because analyzing the relationship of urban quality to place is the key to successful urban planning and design. Its usefulness extends to neighborhoods and cities to understand the functions of social and environmental interactions, etc., with their different dimensions and levels (Garau & Pavan, 2018; Lucas & Romice, 2008).

Thus, the chapter was determined by exploring and clarifying the relationship of urban quality to place. Several steps were adopted. The first step aims to define the concept of urban quality by addressing the concept of quality of the urban environment, urban quality historically, the intellectual traditions of the development of the concept, and its basic goals and principles. The second step aims to review and discuss a group of previous studies, which focused on clarifying the relationship of urban quality to the place. Finally, the third step of discussing the main research axes leads to the research conclusions.

2. Concepts of Urban Quality

This section is organized as follows: first, the general concept of quality in urban environmental studies, second, taking note of both the history of the concept of urban quality and the changing nature and scope of this concept over the years, and third, the intellectual traditions of the development of the concept, and finally identifying the design goals and principles to the concept of urban quality.

2.1. Concept of Quality in Urban Environmental

Quality in urban environment studies is the focus of many urban planners and designers, each conveying their conceptual views and transforming them into themes. These views depend heavily on the urban context, in both time and place. The concept of quality has been mentioned in several contexts, including:

- The concept was put forward within the urban planning framework, as it serves as a basic methodology for urban strategic planning. It includes three types (design, redesign, and compatibility and performance), and the three types of quality must be understood, as well as the relationship between quality and productivity and the benefits of quality improvement to managing quality to achieve the purpose of which (Gülersoy et al., 2009).
- The concept was used within the framework of performance[1] studies, where performance is directly related to the quality of the urban environment. The word acquires a professional meaning in the terminology of the construction industry, i.e., the behavior associated with use, and that the quality of the product is a sign of performance, or that the performance of the product is a sign of its quality, and this confirms the interdependence of the two concepts, and the direct link between performance and the quality of the urban environment (Gülersoy et al., 2009; Rouli and Kusumastuti, 2020).

Quality is understood in two separate meanings. The first meaning is the characteristic of a person or thing that relates it to nature or defines its quality within a particular category, such as Kind and Type. The second meaning is related to the context of the level of physical quality, such as condition, value, rank, degree, measure, and distinction. Quality was used as a concept that creates socially beneficial places and promotes positive behavior, in addition to the importance of quality in contributing to the city's competitiveness and reducing environmental damage (Gülersoy et al., 2009).

- On the part of the individual's definition of high quality in urban environments, it depends on individual perceptions, which may differ as a result of being affected by gender, age, religion, and culture because what constitutes it is a personal matter to some extent, as quality is not an inherent feature of the environment, but rather a function linked to the behavior resulting from the interaction of characteristics of the urban environment and the characteristics of the person. It cannot be assumed that every building, neighborhood,

1 Performance: Is a factor that has a direct impact on the spatial satisfaction of users, and the degree to which the product responds to the needs of the user, and this is what makes quality a relative and subjective concept. User needs are described as the physical, psychological and social conditions that help users perform activities efficiently (Gülersoy et al., 2009).

town, or city user has the same attitudes toward the urban environment (Adams, 2014; Talen, 2002).

- Finally, measuring the high quality of urban environments is not based on the subjective opinions of users/residents, their level of satisfaction with a service or commodity, or their perceived evaluation of it. Still, the quality of the urban environment can also be measured objectively through evaluations of building construction, availability of green spaces, and more (Dempsey, 2008).

Deduction 1

Quality in the urban environment is a complex concept (relative and subjective), generally based on personal perceptions. High quality is achieved when the environments are rich in functional, economic, social, cultural, aesthetic, and psychological values. This creates beneficial places socially, economically, and environmentally, gives the greatest value to the user, and achieves his well-being.

For a more accurate understanding of the subject, the concept of urban quality will be dealt with in the sequence of the historical presentation of the concept in the next paragraph.

2.2. Historical Revision of The Urban Quality Concept

The concept of urban quality has been used in a variety of synonyms such as: "good city shape" (Lynch, 1981), "environmental quality" (Rapoport, 1983; Kamp et al., 2003), "characteristics of a good city" (Jacobs and Appleyard, 1987; Jacobs, 2011), and "spatial quality" (Moulaert et al., 2011), ...etc. Sholihah, 2016; and Bieda & Telega, 2021 suggested that urban quality can be defined through a set of terms or concepts and their multiple interpretations within the trends of many urban theorists, planners, and designers. They could be put forward historically as follows:

- Howard, 1898 invented the theory of the (Garden City), which was a response to the poor quality of urban environments due to the industrial age in Britain, giving residents wide access to open green spaces and large quantities of fresh air. At the same time, in the United States, this preoccupation with the quality of urban environments was permeated by the short-lived but influential City Beautiful movement, articulated in Burnham's Plan of Chicago, whose aim was to restore lost visual and aesthetic harmony to the city (Dempsey, 2008).

- Jacobs, 1961 was considered the first to discover urban quality from the hypothesis of activity that produces and reflects the quality in the urban environment, as she referred to the death and life of major American cities and shed light and drew attention to the loss of those cities' vitality as they are not vibrant. She emphasized several key elements of quality, which include safety, a mix of uses, public connectivity, and diversity of components, with four conditions represented by multiple uses, low-rise buildings, variation in building age, and sufficient density (Sholihah, 2016).
- The quality related to the natural environment was used considering the physical and chemical standards, and they were adopted as international rules according to which countries are formed. In 1978, the Chilean government followed the environmental controls for water, air, and activities related to nature and man, and the United Kingdom Ministry of Environment published the first indicators of sustainable development. Nakagushi, 1980 used local environmental indicators for the cities of Japan. In Argentina in 1991, important indicators were identified in the structure of basic infrastructure (Sarmento et al., 2005; Marans, 2012).
- Lynch, 1981 formulated five dimensions of city performance in the "good city theory," which produces a good city shape and the desirable qualities of a successful urban space, represented by dynamism, sensibility, adequacy, accessibility, and control. He considered that urban quality is due to the joint effect of the place and the society that occupies it, and that performance can be measured only by referring to the spatial form of the city (Gülersoy et al., 2009; Abbas & Kawther, 2014).
- Jacobs and Appleyard, 1987 suggested an "Urban Design Statement" intending to improve the quality of the future urban environment in response to the challenges of modern urban design through indicators such as (livability, identity, centrality, access to opportunity, imagination, and joy, authenticity, meaning, communication and public life, urban self-reliance, and justice). They also emphasized the importance of livable streets and neighborhoods with inclusive physical characteristics under living standards (Sholihah, 2016).

- Tibbalds et al., 1993 published a report on issues of urban quality, including eight elements that were expressed as factors for the quality of the urban environment in London, which include human scale, clarity of structure and identity, visual enrichment, urban management, mixed-use, public and private spaces, safety and cleanliness, and pedestrian (Chapman & Larkham, 1999).
- Smith et al., 1997 developed the principles of quality and the factors that must be available in the urban environment: livability, personality, communication, mobility, personal freedom, and diversity (Adams, 2014).
- Perfect and Power, 1997 emphasized in their book "Planning for Urban Quality" that the elements of quality in urban environments may come from a group of factors related to a sense of place, such as clarity, collective memory, issues of historical continuity, and diversity in a multicultural society. They argued that these factors are crucial to creating high-quality urban spaces, and as a result, the quality of spaces is reflected in the quality of life (Chapman & Larkham, 1999).
- Montgomery, 1998 added that the concept of urban quality, represented by a sense of place, is linked to the social, psychological, and cultural dimensions of the place, and suggested three indicators of quality, which constitute a successful place, represented by activity, form, and meaning, with secondary indicators for each of them.
- Florida, 2002 in proposing the term "creative economy and creative class," he explained that six quality points in the urban environment create a creative place, including diversity, social interaction, authenticity, lifestyle, identity, and creativity. The quality Florida defines is specific and entails a combination of factors that together make the city an attractive place for the creative class to reside.
- Hastings et al., 2005 claimed that the current focus on the quality of the urban environment is stronger than ever, particularly policy on livability and the provision of clean, safe, and green public spaces and streets (Dempsey, 2008).
- Durmaz, 2012 summarized the urban environment quality indicators formulated by Landry as economic vitality, social vitality, environmental vitality, and cultural vitality. Hall, 2012 revealed the quality of the urban street through individual experiences, as the

author clarified three main indicators in assessing the quality of the urban street represented by the ability to adapt, flexibility, economic and social complexity, diversity, longevity, and vitality (Sholihah, 2016). These traditions continued to support planning policies and guidelines throughout the twentieth and twenty-first centuries by defining the minimum housing density and recommending high-quality housing designs and urban planning for the importance of design and quality in attracting people back to urban neighborhoods, with the need to focus on good quality public spaces and attractive homes, and this has led to a large number of mandatory urban design guidelines (Dempsey, 2008).

The urban quality indicators and their interlinkages are presented in Figure 1.

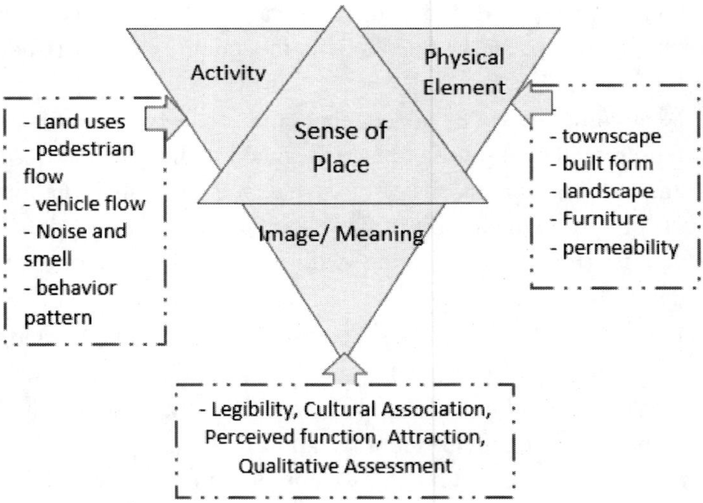

Source: author.

Figure 1. Urban quality indicator.

Deduction 2

Therefore, it can be said that urban quality is a complex concept that differs according to the opinions of theorists, designers, and recipients. For example, urban quality for Montgomery (1998) is seen from a much broader perspective than the physical characteristics of buildings, street patterns, and spaces, while Lynch (1981) defines it as the influences of the relationship between a place and the community that occupies it. The focus of the literature was on defining the concept of urban quality by presenting indicators, which vary between the elements and factors of quality in the urban environment and environmental quality and are directly or indirectly related to the social, economic, environmental, cultural, and even psychological dimensions of the place, as can be seen in Table 1.

Table 1. Scheme of the concept of urban quality historically, according to urban theorists and designers

Theorists	Year	Theory
Howard	1898	Providing healthy, clean, high-quality urban environments.
Jacobs	1961	Determine the elements of urban quality (safety, mix of uses, public connectivity, and diversity of possibilities).
Porteous & Nakagushi	1991-1971	Use environmental quality related to the natural environment, involving perceptions, attitudes, and subjective values.
Lynch	1981	Determining five dimensions of city performance that produce a good urban form, represented by vitality, accessibility, Fit, sensation, and control, depending on the spatial form of the city.
Jacobs & Appleyard	1987	Improving urban environment quality through livability, identity, authenticity and meaning, communication, self-reliance, justice, access to opportunities, imagination, and fun.
Tibbalds	1993	Determining urban environment quality factors (human scale, clarity of structure and identity, visual enrichment, urban management, mixed-use, public spaces, and private places, safety and cleanliness, pedestrian streets)
Smith et al.	1997	Define the principles of quality for the urban environment (livability, personality, connectivity, mobility, personal freedom, and diversity).
Parfect & Power	1997	Determine urban quality elements - sense of place factors (clarity, collective memory, historical continuity, diversity in a pluralistic society); thus, the quality of places reflects the quality of life.
Montgomery	1998	Suggest indicators of urban quality, represented by the sense of place (form, activity, and meaning), associated with the place's social, psychological, and cultural dimensions.
Florida	2002	Determine urban environment quality factors that create a creative place (diversity, social interaction, authenticity, lifestyle, identity, and creativity)
Hastings et al.	2005	Focusing on urban environment quality through (the possibility of living, providing clean, safe, and green public spaces and streets).
Durmaz & Hall	2012	Urban quality indicators in attracting people to urban neighborhoods (economic vitality, social vitality, environmental vitality, and cultural vitality). Also, revealed indicators of urban street quality, adaptability (flexibility, economic and social complexity, and diversity), longevity, and vitality.

Source: authors.

2.3. The Intellectual Traditions of the Evolution of the Urban Quality Concept

Heath et al., 2003 identified three intellectual traditions, allowing for a discussion of the possibility of providing high-quality urban environments and delineating the sequential conceptual process of the theoretical evolution of city design and planning over the past century. The author claimed that no tradition is more important than the other or replaces it. These traditions will be clarified in subsequent paragraphs (Dempsey, 2008; Abbawi & Abdul-Kareem, 2016).

2.3.1. Visual-Artistic Tradition

This tradition focuses on the visual form of the urban environment through the visual experience of buildings and spaces. It is a well-known early intellectual school in architecture, especially in urban design. It used the oriented visual-artistic concept of the product, which focuses on the visual characteristics and aesthetic experience of urban spaces more than its focus on the cultural, social, economic, political, and spatial elements and processes that contribute to the success of urban spaces.

In the late nineteenth century, Sitte, 1889 argued that space should be organized according to visual experience and that only what the viewer sees has artistic significance. Sitte's views influenced twentieth-century planners and architects, notably Raymond Unwin and Gordon Cullen (Collins, 2006). Unwin said that the lack of aesthetics and comforts of life forces us to admit that our city-building work in the last century was not well done (Unwin, 1994). In the post of WWII period, design focused on the appearance and layout of the urban environment. However, Abercrombie, 1998 stated that beauty should rank first in the list of beauty, health, and comfort, which being the quality that must pass through the whole to uplift and achieve the dignity and well-being of city life (Abercrombie, 1998).

Cullen, 1961 provided influential guidelines on the townscape. He stated that a successful city must be designed with people's visual experience in mind because the perception of the environment entirely is almost through vision. He introduced the concept of serial vision and emphasized that the urban environment should be designed from the point of view of the moving person, so the environments are tested as a dynamic time sequence to describe the visual aspect of the city and the kinetic aesthetic experience in the city is the most important part of the visual tradition (Saikia, 2013).

Criticism of beauty-driven urban planning and design spread especially among those adhering to the city's pragmatic traditions. Robert Owen, Lever Brothers, and Joseph Rowntree argued that providing clean and healthy living conditions for workers and residents was the key to solving social problems. It cannot be said that the artistic visual tradition was no longer followed when the tradition of social use emerged, but rather, this school of thought continues to this day. There is a group of visual-artistic features of the urban quality, which can be listed as follows:

I. Visual attractiveness is an advantage of the quality of the urban environment, which positively affects the quality of life and the residents' satisfaction with the place. It contributes to the residents' feeling of connection and belonging to a place (Dempsey, 2008).
II. What is attractive to one person may not be to another. However, there seems to be an implicit assumption that the term is comprehensible to all and needs no definition (Ubaid & AL-Ataabi, 2022).
III. It is appropriate to combine objective and subjective indicators to measure the attractiveness of the urban environment (Sitte, 1968).
IV. The visual complexity of the urban environment is described as a characteristic of high-quality urban environments in which there must be surprising visual contrasts (Alkurukchi, 2012).
V. The perceived visual experience of the urban environment creates a character in a place that comes from the importance of the local context. The visual experience of the urban environment directly impacts people's sense of a particular place, making them feel safe and comfortable (Cullen, 1961).

Deduction 3

From what was presented above, it could be concluded that the importance of organizing and visual cohesion of the components of the urban environment to create successful and high-quality urban spaces that respond and contribute positively to their context and provide visual comfort.

2.3.2. Social Usage Tradition

This tradition focuses on the individual's experience in urban environments (the individual's social activities in urban places). This school of thought was associated with the school of the visual-artistic tradition through the visual perception of the urban environment. It moved towards the general perception of the urban environment, focusing on how people use and occupy space (Dempsey, 2008; Ali et al., 2003).

Abbawi & Abdul-Kareem, 2016 pointed out that Lynch, 1960 pioneered a new trend. He mentioned that every citizen has long associations with a part of his city. Lynch image is steeped in memories and meanings. Other theorists point out that the design and shape of the city must support the necessary and non-essential factors for daily life in the city. On the other hand, Gehl (1971) believes that places have meaning because of the activities that take place in them.

The advantages of social use of urban quality can be listed as follows:

I. A well-connected and permeable urban environment is described as a characteristic of high urban quality and is physically represented as small urban blocks. Gehl (1971) argues that well-connected pedestrian roads are well-used, preferable to deserted or quiet roads, and small block sizes provide pedestrians with more alternatives to the track, allowing them to navigate easily. In addition to fair access to urban resources of social value to citizens, this is known as social justice.

II. Few guidelines were provided on the size of small urban agglomerations. Burton and Mitchell advised that the blocks be of varying short lengths (60-100m) to allow for diversity (Burton & Mitchell, 2006).

III. Mixed use in the neighborhood provides users with services and facilities that are regularly needed in their immediate surroundings, as Jacobs explained that there is a need for a mix of uses to express and support many people (Jacobs, 1993).

IV. Visual clarity or visual quality, according to Lynch, 1962, is the ease with which parts of the city can be identified and can be organized in a coherent pattern. A well-defined city is one whose areas, paths, borders, and transition points can be distinguished easily. The elements are interconnected and integrated in one clear formation. Visual clarity is related to transparency, representing the degree to which people can see the edge of the street, public space, and human activity.

V. Safety is an important element in a high-quality urban environment. Al-Kaissi & Al-Askary, 2016 explained that when people can understand the layout of the place and what is happening in the place improve the safety of community. Urban masses, activities, physical elements of the urban environment, urban planning, and building height and shape can all contribute to the user's sense of safety and comfort and lower crime levels. This confirms that safety is an essential feature of high-quality urban environments.

VI. Kadhim & Shok, 2021 referred to Jacobs, 1961 and Alexander, 1977 who described the character of the place, which creates a sense of the components of the place, as an essential feature of the urban quality, resulting from a combination of the features of the urban environment and the social life in the neighborhood, and thus the perceived character of the place could contribute to the feeling of user's safety when navigating the streets, perhaps as a result of natural, non-coercive observation. Measuring personality relies on residents' subjective opinions to successfully capture the essence of the local context.

Deduction 4

Here, it turns out that individuals within the community need to work together and interact for the sustainability of society. The concept of community is fundamental to people's general quality of life, a sense of belonging, achieving social cohesion, a sense of place, and creating a livable environment, which constitutes a high-quality urban environment.

2.3.3. Making Places Tradition

This tradition focused on the idea of spaces becoming places because the urban environment supports the activities in urban spaces (Nasution & Zahrah, 2014). This tradition has recently resulted from the merging of previous traditions, as the concept of urban design has developed over the past twenty years to provide places for people. Buchanan, 1988 indicated that urban design was an essential basis for making a place, as places are not only specific spaces but are those events and activities that make them possible and meaningful (Heath et al., 2003; Hunt, 2008).

Interpretations and ideas of schools of thought (visual-artistic and social use) continue to be accepted by current theorists and practitioners and then incorporated into a third tradition, which is the making of places. The supporters of this tradition are interested in designing the urban environment as an aesthetic entity and behavioral framework that emphasizes the diversity and activity that help create successful urban places, and this affects the quality of the physical environment that supports the functions and activities that occur there (Alkinani et al., 2018). The tradition of place-making is closely related to the theoretical model of sustainability, namely that the urban environment provides a physical and social environment for current users and their collective needs, and all future residents, users, and their collective needs. It is arguable that for the urban environment to be of high quality, it must be sustainable, as the sustainable urban form is closely aligned with theoretical and political interpretations of high-quality places (Abbawi & Abdul-Kareem, 2016).

The concept of place-making represents the active participation of the community, at different levels, in planning and managing places to achieve safe, vibrant, welcoming, and attractive places for people, and fit for society in the long term, and thus achieve well-being (Hunt, 2008; Petrovič & Murgaš, 2021). Place-making is the process of enrichment with high-quality urban details, principles, and elements to create preferred and popular urban living environments for people to build resources, share ideas and knowledge, and preserve social, economic, and environmental values (Dempsey, 2008).

As for the features of place-making of high urban quality, they can be summarized as follows[2]:

I. Neighborhood accessibility describes the ease of residents to access services and facilities at a reasonable cost, time, and ease. A neighborhood with equitable access to services and facilities, employment opportunities, and green spaces, or the means (such as public transportation) to access them, provides an equitable distribution of resources and equality of conditions to support quality of life (Dempsey, 2008).

II. Inclusiveness is a characteristic of the quality of the urban environment, which represents the ability of all people to fulfill their potential without suffering from negative impacts, such as unemployment, low income, poor housing, and poor health. It is conceptually related to equitable access and intergenerational equality, i.e., supporting the needs of the less fortunate in society. The comprehensive environment is comfortable, pedestrian-friendly, and welcoming for all residents, as it welcomes all sectors of society. The comprehensiveness of the public space can be affected by the existing

[2] There is significant overlap in the features emerging from the three traditions of thought. Therefore, coherence, permeability, clarity, security and attractiveness are not discussed.

services and facilities and the quality of those services and facilities. It can directly affect the level of use of the place (Ali & Abdulsahib, 2019).

III. High population density is becoming an increasingly important but controversial feature of sustainable urban environments. However, there is no consensus on whether high density is a positive feature of the urban environment, so it cannot be said that low-density urban development is of poor quality. On the contrary, many have argued that it is a key feature of high-quality neighborhoods as to how high or what it should be. Governments have embraced the principle of increasing residential densities to use land more efficiently to pursue sustainable urban development (Abbasi et al., 2016).

IV. Good maintenance. Recent research has found that the maintenance of infrastructure and public cleanliness were among the main concerns of users, which is reflected by the adoption of governments to an appropriate strategy for living. Maintenance is also related to the quality of service the local authority provides, such as street cleaning, and thus reflects physical environment quality (Baklazhenko et al., 2019).

V. Place-making concept focuses on creating a place more than designing it. Physical things alone do not make a place, but what makes it is how people respond and interact in this place as the target is residents and users, so attention is paid to activities, utilizations, and non-material factors such as identity, cultural, aesthetic, and social needs, as well as meanings and memories more than material factors, which represent the physical form of a place. This can be achieved by respecting local traditions, using local materials, and retaining cultural features. Therefore, identity is always seen as a feature of high-quality urban environments, which reflects the importance of the local context (Burton & Mitchell, 2006; Ahmed, 2019).

Deduction 5

Place-making is understood as the art that is presented to people in the form of urban environments with distinct characteristics. It is a multi-faceted approach that has been adopted in the areas of planning, design, and management of spaces and public places for social purposes aimed at enhancing culture and social relations between members of the local community to obtain luxury and comfort in these places, with supporting to its continuous development to keep pace with the development of society. Creating a vibrant, safe, and attractive place is the key to success for the city's and residential neighborhoods' economic, social, environmental, and cultural well-being.

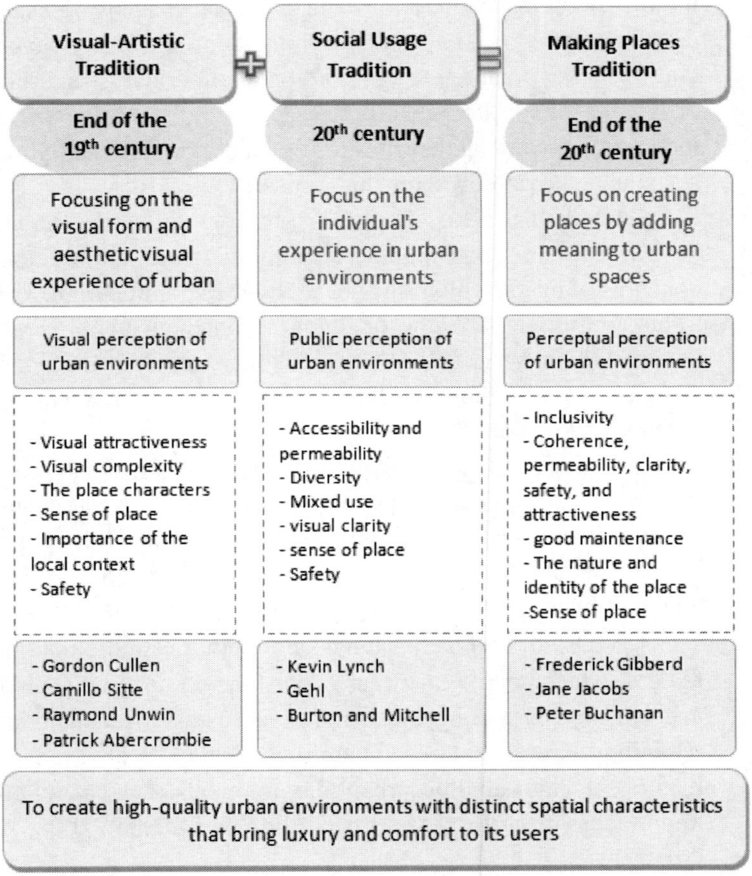

Source: author.

Figure 2. The intellectual traditions of the development of the concept of urban quality.

It is clear from the previous that the focus of the visual-artistic tradition at the end of the 19th century was on the visual form and the aesthetic visual experience of urban environments, which achieves visual perception through a set of features represented by visual appeal, visual complexity, the character of a place, sense of place, the importance of local context, and safety, while the tradition of social use in the 20th century focused on the individual's experience in urban environments, which achieves general awareness through a set of features represented by the ease of access and permeability, diversity, mixed-use, visual clarity, sense of place, and safety. The tradition of place-making at the end of the 20th century focused on creating places by adding meaning to spaces and achieving the sensory perception of urban environments through a set of features represented by inclusivity, interdependence, permeability, clarity, safety, attractiveness, good maintenance, the nature and identity of the place, and the sense of place. The purpose of these three traditions is to create high-quality urban environments with distinct spatial characteristics that bring their users luxury and comfort, as seen in Figure 2.

2.4. Objectives and Principles of Urban Quality

Studies have identified many goals for urban quality. Carmona, 2019 focused on urban quality's health, social, economic, and environmental goals. Garau et al., 2018 mentioned three basic goals for urban quality: improving the quality of life, providing better services to citizens, and creating an environment that promotes innovative jobs (Garau & Pavan, 2018). The Urban Design Protocol (2005) was developed for New Zealand and was prepared by the Ministry of the Environment in partnership with the Urban Design Consultative Group to clarify urban quality goals, which include protecting heritage and cultural identity, improving economic performance, increasing economic value with higher returns on investment, reducing maintenance and management costs, creating more productive workplaces, creating productive, strong and attractive urban environments, creating interconnected and accessible places while providing a mix of services that users need, enhancing safety and reducing crime, improve psychosocial well-being by increasing opportunities for physical activity and encouraging healthy lifestyles, and creating high-quality urban environments which require the collaboration of a wide range of disciplines, sectors, and users due to the complexity of cities, neighborhoods, and buildings (Ministry of Environment – New Zealand, 2005).

Elsemary, 2020 stated that urban quality at the level of pedestrian paths is considered a basic building block for constructing high-quality urban environments, achieving several goals, including increasing health well-being, achieving physical well-being, enhancing social well-being, achieving work quality, improving psychological state, and finally high-quality pedestrian paths that reduce traffic congestion and achieve security and safety (Hesham & Elsemary, 2020; Dodge et al., 2012).

The studies put forward several design principles for urban quality, which can be summarized as diversity, which ensures the availability of opportunities for all, creativity, that adds richness and diversity, provides vitality, and facilitates the use of new ways of thinking and preparation to solve the problems of the urban environment, harnessing technology and envisioning a new future, where the creative urban environment supports dynamic urban cultural life and enhances urban identity.

Connectedness, which promotes choice, supports social cohesion, makes places lively and safe, and facilitates contact between people, character, as it ensures that the new buildings and spaces are distinctive, appropriate to their location, and encourage their historical identity, which gives value to cities and creates a distinctive place, and encourages tourism and investment, which creates a sense of belonging to the place, clarity, by securing visual contact, the openness of vision along the urban landscape, achieving a sense of security through a sense of control over the place, and collaboration between all decision-makers (central and local governments, professionals, developers, and users).

Aesthetics, where the urban environment with its aesthetic and artistic characteristics enhances the element of exploration and prevents a sense of boredom and monotony, and thus supports the role of places as an element of attraction for people and their activities, and flexibility and adaptation, where are variables that affect the relationship between objective and subjective conditions in the urban environment. Adaptation is at the level of individuals among themselves, the adaptation of individuals to situations and external conditions, and the adaptation to the urban environment (ministry of Environment-New Zealand, 2005; Adams, 2014; Garau & Pavan, 2018; Hoseini & Mokhtari, 2013).

Deduction 6

Urban quality has main goals that represent health, economic, social, and environmental goals, in addition to achieving the well-being of the place in each of these aspects to improve the quality of life by creating safe, attractive, and creative urban environments.

Regarding the concept of urban quality, urban environments are considered to have the ability to achieve diversity, communication, and interaction greatly among people of different ages, cultures, and social levels. The principles of urban quality are summarized in what aims to achieve a better and more efficient life, address challenges, focus on innovation, create opportunities, and achieve public welfare.

3. A Critical Review on Urban Quality Correlated to Place

In this section, the focus will be on reviewing and discussing a group of the most important previous studies that dealt with the issue of urban quality by focusing on the relationship with a place within its various factors, characteristics, and dimensions.

3.1. Urban Quality-Space Correlation Based on Deterioration of Historical Urban – Case Study in Istanbul, Turkey

Gülersoy et al., 2009 study dealt with place's relationship with urban quality by introducing a strategic approach to plan the quality of deteriorating historical urban areas at the social, cultural, functional, economic, spatial, and structural levels to explore urban quality in comprehensive quality management. The study was based on a research project entitled "Improving the Quality of the Environment in the Urban Fabric: A "Strategic Planning Model for Quality" conducted at Istanbul Technical University in Turkey in 2005. The study indicated the need to vitally revise the quality of deteriorating urban areas due to urban growth problems resulting from the rapid population increase and the dynamics of urbanization locally. These generated fears of Turkey's inability to adapt to urbanization. Hence, the study focused on the eight-stage strategic quality plan represented by establishing a strategic planning team, analyzing the current situation, defining the mission, defining the vision, defining goals and strategic objectives, developing basic strategies, preparing strategic plans and projects, and monitoring and evaluation.

This plan indicated the inability of physical interventions alone to create a sustainable and effective environment. Therefore, it stressed the necessity of integrating the spatial, economic, social, functional, cultural, and structural dimensions to create a positive environment for urban quality, encourage active quality management, provide a healthy and comfortable environment, and encourage optimal communication, functionally preserve the historic environment, provide optimal cost and financial support, ensure flexibility and applicability, support community participation, provide social and cultural cohesion, providing equal access to the community, and providing a legal and administrative framework integrated with planning and preservation processes (Gülersoy et al., 2009).

3.2. Urban Quality-Place Correlation Based on Social, Behavioral and Material Features

Hoseini & Mokhtari, 2013 clarified the relationship of the place with the issue of urban quality by analyzing and evaluating its social, behavioral, and material features, which affect the quality of public places and urban life, focusing on creating a sense of place and belonging to it. The study aimed to compare the social and special features of successful public places, the relationship between people and specific spaces, people's behavior, needs, and necessities, and analyze and evaluate the features affecting the quality and success of public places. The study referred to the spatial advantages of achieving quality in public places, which represent the diversity of activities and spaces and the provision of flexibility, security and safety, ease of access, and luxury. The study focused on the main features of successful, high-quality public spaces, which include accessibility and communication, which means easy access to public spaces; user and activity, in terms of people collaborating and engaging in public space activities; comfort and perception, which refers to feeling comfortable in a space and creating a good mental image of public space; and the sociability[3] of public space, which means a place where people come together to meet each other (Hoseini & Mokhtari, 2013).

3.3. Urban Quality-Space Correlation Based on Infrastructure and Buildings. Case Study in Zurich

Brici, 2014 developed a framework and validate it through three development areas in the city of Zurich. He clarified the relationship of place with the issue of urban quality by assigning evaluation indicators for the place factors

3 Sociability: social mixing in public places (Hoseini & Mokhtari, 2013).

associated with urban quality, which includes the tangible material factors represented by the spatial aspects (infrastructure, buildings, and public places). In addition to the intangible economic and social aspects (jobs, demography, density), and the non-material factors represented by (cognition, memory, and identity).

Brici, 2014 stated that urban quality is a complex concept and that it is necessary to understand its various aspects, and its indicators must be broad enough to include all these relevant aspects. Because of the complexity of the term because of its different aspects, the study noted that there is no specific official definition. It must be divided into specific levels to make the concept more understandable and usable. For example, the urban quality is divided into buildings, alleys, and public places. He focused on the place factors associated with urban quality, which include:

- centrality related to (logistical centralization, functional centralization, daily life centrality, symbolic centralization),
- accessibility (public access, porosity, rights of use),
- ease of use represented by (the ease of multiple uses, i.e., the ability of the space or building structure to be used for several purposes or to receive new uses),
- the ability to adapt, which means (the general flexibility of buildings, and the flexibility of the ground floor), the diversity that includes (functional diversity, social diversity, property diversity, and spatial diversity),
- interaction, that indicates (social density, interaction spaces),
- personalization (identity, use of personalization), and the design, related to (architectural competitions, architectural quality).

3.4. Urban Quality-Space Correlation Based on Physical- Factors Measures~~Factors Measures~~

Fachrudin, 2020 deliberated on the Urban quality measurement and its influence on sense of place. He clarified the relationship of place with the subject of urban quality through physical factors (functional use and visual and urban experience) and cognitive and perceptual factors (identity, structure, and meaning of place) affecting the sense of place and urban quality. The study stated that a city is a living place that must have an effective urban quality to be a better attractive place for people and form a sense of place. The quality of urban areas is paramount and must be considered when creating a sense of place. The study showed that urban quality consists of two factors: the physical characteristics of the material preparation factor and the cognitive and perceptual factors. The concentration was on the physical features, represented by functional use (buildings and ease of access), visual experience (building facades, human scale, design of landscape, and street furniture), urban experience (diversity of activities and good pedestrian path), and considered as independent variables. In contrast, the sense of place consists of identity, structure, and meaning, representing the dependent variable. Quality can be measured by people's perception of an object based on proportion, size, shape, sense of direction, and the nature of communication between spaces. As for the sense of place, it can result from the identity related to the shape and history of the place, and it contains some elements represented by scale, size, proportion, diversity, color, texture, decorations, smell, sound, the integrity of history, and others (Fachrudin, 2020).

3.5. Urban Quality-Space Correlation Based on SustainabilityCase Study in Indonesia

Sholihah, 2016 presented an overview of urban quality indicators towards a sustainable and sophisticated urban life. The study adopted Indonesia to clarify the relationship of place to the issue of urban quality by presenting urban quality assessment indicators related to the place at the formal, social, cultural, functional, visual, and sensory levels. The aim of this study was represented by identifying the elements related to place, which are greatly associated with urban quality, to achieve a sustainable, high-standard life. The study stated that urban quality is a vague concept and that the associated indicators must be known, and this vagueness is due to their diversity and complexity and not because of their lack of content. Some literature on urban quality was reviewed regarding urban quality assessment indicators. After the analysis process, the study presented urban quality indicators within the formal, functional, social, cultural, visual, and sensory dimensions. These include clarity, dynamism, accessibility, walkability, diversity, connectivity, adaptability, creativity, form and visual quality, transparency, security, visualization, and originality (Sholihah, 2016).

3.6. Urban Quality-Space Correlation Based on Occupants Satisfaction

Bonaiuto & Fornara, 2017 considered the residential satisfaction and perceived urban quality and analyzed the relationship of place with the issue of urban quality by addressing the concept of satisfaction with place and satisfaction with residential neighborhoods, particularly within the psychological dimension, and considering it as a comprehensive assessment of the quality of the residential urban environment. The study showed that urban quality is multi-dimensional, and the quality of the residential urban environment may include architectural and urban features in terms of spatial features related to architectural and city planning, including green spaces; human and social characteristics, which are represented by the population and the type of social relations; Functional features, associated with the available services; Contextual features, represented by lifestyle, pollution, health, maintenance, and care. The study identified a list of indicators of the perceived quality of the residential environment that achieve the satisfaction or dissatisfaction of the residents, which are: (building aesthetics, building density, building size, internal practicality, external connections (in the city), green spaces (presence and care), security, tolerance, and sociability. Friendliness, school services, social welfare services, sports services, socio-cultural activities, business services, transport services, relaxation versus the painful pace of life, an unpolluted environment, maintenance, and care (Bonaiuto & Fornara, 2017).

3.7. Urban-Space Quality Based on Geographical Approach

Petroviˇ & Murgaš, 2021 described the relationship between urban, space and quality of urban life based on the geographical approach. The study dealt with place's relationship with urban quality by examining the relationship between urban space (place) and the quality of urban life by addressing the concept of livability and achieving satisfaction with place (the population's satisfaction with their lives). This study aimed to find out the relationship between urban space (considered as a place) and the quality of urban life, to transform urban space into a measurable form and define it in a tangible form, thus providing decision-makers of public policy with a valid basis for decision-making. The results contribute not only to urban and geographical knowledge but also to improving urban planning and public policy in general. The study defines livability as a comprehensive expression of the conditions of a place to live in, the degree to which the place is suitable to live in and describes the overall conditions for a decent life for all residents of cities, regions, and societies, including their physical and mental health. It is also defined as the degree to which a species' adaptive living environment is suitable.

When applied to human society, it indicates the suitability of institutional arrangements for needs. Thus, livability was defined as the level of the external conditions of a place for a good life. The quality of life, either urban or rural, is according to where the person lives. The urban quality of life is defined as the living conditions in urban areas, especially cities. It has two dimensions: the first is subjective, represented by satisfaction with life, and the second is objective, related to living conditions. Thus, urban quality of life is defined as an expression of satisfaction with our life in the city. It is often identified with well-being, life satisfaction, or happiness and is an important factor in a city's attractiveness or growth. Individuals express life satisfaction, place satisfaction, happiness, and other variables in different measures. The study concluded that livability is an objective dimension of the quality of urban life, that is, the quality of the place (Petroviˇ & Murgaš, 2021).

Deduction 7

It is clear that the issue of urban quality and its relationship with place is addressed through several aspects, including; Integration of spatial, economic, functional, social, cultural, and structural dimensions to create a sustainable environment (positive for urban quality); Spatial characteristics (social, behavioral, and physical) influencing the quality of places and urban life; Spatial factors (tangible and intangible physical factors, and non-physical factors), as indicators of urban quality assessment, as well as physical factors (the physical characteristics with cognitive and perceptual factors of place affecting urban quality and sense of place.

The formal, social, cultural, functional, visual and sensual dimensions of place are indicators of urban quality assessment; Spatial, social, functional, and contextual dimensions are urban quality dimensions and/or indicators of the residential environment for achieving satisfaction with place. Finally, the relationship of space with the quality of urban life through the concept of livability and satisfaction with place, which represents an objective dimension of the quality of urban life, that is, the quality of place. Table 2 summarizes the important factors on the role of space in the concept of urban quality.

Table 2. The most important terms covered by studies on the role of place in urban quality

previous studies	Primary and secondary vocabulary	
Gülersoy and other, 2009	A strategic approach to planning the quality of declining historic urban areas.	Creating a sustainable environment is a positive issue for urban quality by integrating the economic, functional, social, cultural, structural, and spatial dimensions.
Hoseini & Mokhtari, 2013	Spatial characteristics (social, behavioral, and physical) affecting the quality of urban places and urban life.	- Access and communication / user and activity /rest and imagination / sociability of public space. - Diversity / flexibility / security and safety / luxury.
Fachrudin, 2015	Physical factors, and cognitive perceptual factors of place affecting urban quality and sense of place.	- Physical factors: Physical properties (functional use / visual experience / urban experience). - Cognitive and perceptual factors (structure, identity, and meaning).
Brici, 2014	Place factors for urban quality (tangible and intangible physical factors, and non-material factors).	Centralization / Accessibility / Ease of Use / Adaptability / Diversity / Interactivity / Customization / Design.
Sholihah, 2016	Indicators of urban quality assessment related to place within the formal, social, cultural, functional, visual, and sensory dimensions.	Clarity, vitality, accessibility, walkability, diversity, connectivity, adaptability, creativity, the form and visual quality, transparency, safety, imagination, and originality.
Bonaiuto and Fornara, 2017	Urban quality indicators: it is important for the residential environment by achieving satisfaction with the place.	- Dimensions of urban quality for the residential environment are spatial, social, functional, and contextual dimensions. - Indicators of the quality are aesthetics, density, size, internal practicality, external connections, green spaces, security and tolerance, sociability and friendliness, services, relaxation, unpolluted environment, maintenance, and care.
Petroviˇc & Murgaš, 2021	The relationship of urban space (place) with urban quality of life by addressing the concept of livability and achieving satisfaction with place.	- Perception of urban space and urban quality of life, through livability and satisfaction with place, or satisfaction with the quality of life, the physical space. - Livability is an objective dimension of urban quality of life, that is, the quality of space.

Source: author.

Conclusion

The chapter discusses urban quality, which is considered one of the characteristics of contemporary urban environments to promote a sustainable future. Urban quality plays a role in sustainable development and the establishment of contemporary cities in the light of exploring the concept of urban quality and its relationship to place. Various dimensions, factors, and characteristics are essential for a better understanding of the concept of urban quality, the design of high-quality urban environments, and the reconfiguration of built environments to accommodate new urban conditions.

The chapter presented a framework that supports the relationship of urban quality to the place. Quality in the urban environment is a complex concept and generally based on personal perceptions. High quality is achieved when the environment is rich in functional, economic, social, cultural, aesthetic, and psychological values. This creates beneficial places socially, economically, and environmentally and gives the most value to the user. Quality of the urban is a dynamic concept that changes with the introduction of new models and methods. The elements and factors of quality are directly or indirectly linked to the social, economic, environmental, cultural, and even psychological dimensions of the place to create high-quality urban environments with distinct spatial characteristics that bring luxury and comfort to its users.

A review of previous studies concludes three groups of factors for urban and place quality. First, spatial factors include integrating tangible and intangible physical factors, non-physical cognitive, and perceptual factors. Second, spatial characteristics like social, behavioral, and material. Third, spatial dimensions include formal, social, cultural, functional, visual, and sensory.

References

Abbas, S. S., & Kawther, K. K. (2014). The Visual Image Quality for the Future City. *Engineering and Technology Journal*, 33(1A): 1–36. https://doi.org/10.30684/etj.33.1A.20.

Abbasi, A., Alalouch, C., & Bramley, G. (2016). Open Space Quality in Deprived Urban Areas: User Perspective and Use Pattern. *Procedia - Social and Behavioral Sciences*, 216(6): 194–205. https://doi.org/10.1016/j.sbspro.2015.12.028.

Abbawi, R. F., & Abdul-Kareem, B. (2016). Making Healthy place As Contemporary Urban Trend. *Engineering and Technology Journal*, 34(14A): 566–586. https://doi.org/10.30684/etj.34.14A.1.

Abercrombie, P. (1998). Town and Country Planning. London: Routledge.

Adams, M. (2014). Quality of Urban Spaces and Wellbeing. *Wellbeing and the Environment: Wellbeing: A Complete Reference Guide, Volume II.* Edited by Rachel Cooper, Elizabeth Burton, and Cary L. Cooper. John Wiley & Sons:1-21. https://doi.org/10.1002/9781118539415.wbwell064.

Ahmed, H. I. S. (2019). The Principles of Achieving the Quality of Public Plazas and Squares According to the Concept of Placemaking. *Journal of the Egyptian Society of Engineers*, 58(2): 174-186.

Al-Kaissi, A. O. H., & Al-Askary, A. H. A. A. (2016). Visual integration of Historic Cities Centers - Al Rasheed Street. *The Iraqi Journal of Architecture and Planning*, 15(1): 81-1-7. https://doi.org/10.36041/iqjap.v15i1.382.

Ali, K. I., Al-Chalabi, S. A., & Parsegh, M. F. (2003). Spatial Organization and Social Structure. *The Iraqi Journal of Architecture and Planning*, 3(1): 57-70. https://iqjap.uotechnology.edu.iq/article_169837.html.

Ali, S. A. H., & Abdulsahib, S. S. (2019). Knowledge-Based Urban Development the Impact of Knowledge- Based Urban Development in the Growth of Contemporary Cities. *Journal of Engineering*, 25(1): 1-22. https://doi.org/10.31026/j.eng.2019.01.08.

Alkinani, A. S., Najm, H. H., & Ateaa, E. H. A. (2018). Clarity and Visual Ratios of the Vibrant Place Making. *KnE Engineering*, 3(4): 45-67. http://dx.doi.org/10.18502/keg.v3i4.2161.

Alkurukchi, M. A. M. (2012). The Effect of Spatial and Visual Configuration of Museums in The Formation of Visiting Patterns, A Comparative Study Between Iraqi and International Museum Buildings. *The Iraqi Journal of Architecture and Planning*, 11(25): 127-24. https://iqjap.uotechnology.edu.iq/article_159620.

Baklazhenko, V. V., Vovzhenyak, P. Y., & Pavlovskaya, E. E. (2019). The Major Cities Quality Urban Environment Forming Features. *IOP Conference Series: Materials Science and Engineering*, 698(3): 033004. http://dx.doi.org/10.1088/1757-899X/698/3/033004.

Bieda, A., & Telega, A. (2021). The Analysis of Research Hotspots in the Field of Urban Quality. *Sustainability*, 13(17): 9582. https://doi.org/10.3390/su13179582.

Bonaiuto, M., & Fornara, F. (2017). Residential Satisfaction and Perceived Urban Quality. *Reference Collection in Neuroscience and Biobehavioral Psychology*, Elsevier: 267–272. https://doi.org/10.1016/B978-0-12-809324-5.05698-.

Brici, B. I. (2014). *Developing A Framework to Assess Urban Quality An Approach Validated Through Three Development Areas In The City Of Zurich*. Master thesis, Zurich University, Germany.

Burton, E., & Mitchell, L. (2006). *Inclusive Urban Design: Streets for Life*. Taylor & Francis group. https://doi.org/10.4324/9780080456454.

Collins, C. C., Collins, G. R., and Sitte, C. (2006). *Camillo Sitte: The Birth of Modern City Planning*. Dover Publications: 129-434.

Carmona, M. (2019). Place Value: Place Quality and Its Impact On Health, Social, Economic And Environmental Outcomes. *Journal of Urban Design*, 24(1): 1-48. https://doi.org/10.1080/13574809.2018.1472523.

Heath, T., Oc, T., & Tiesdell, S. (2003). *Public Places - Urban Spaces*. 2nd edition, Routledge, London. https://doi.org/10.4324/9781856179041.

Chapman, D. W., & Larkham, P. J. (1999). Urban Design, Urban Quality and the Quality of Life: Reviewing the Department of the Environment's Urban Design Campaign. *Journal of Urban Design*, 4(2): 211–232. https://doi.org/10.1080/13574809908724447.

Cullen, G. (1961). *TOWNSCAPE*. Routledge, London. https://doi.org/10.4324/9780080502816.

Dempsey, N. (2008). Quality of the Built Environment in Urban Neighbourhoods. *Planning Practice & Research*, 23(2): 249–264. https://doi.org/10.1080/02697450802327198.

Dodge, R., Daly, A., Huyton, J., & Sanders, L. (2012). The Challenge of Defining Wellbeing. *International Journal of Wellbeing*, 2(3): 222–235. http://dx.doi.org/10.5502/ijw.v2i3.4.

Hesham, E. M. & Elsemary, Y. M. (2020). Towards Activating the Quality of Life Through Non-Motorized Movement Paths. *Journal of Urban Research*, 36(1): 12-30. https://dx.doi.org/10.21608/jur.2020.90089.

Fachrudin, H. T. (2020). Urban Quality Measurement and It's Influence to Sense Of Place. *IOP Conference Series: Earth and Environmental Science*, 452: 012139. https://doi.org/10.1088/1755-1315/452/1/012139.

Garau, C., & Pavan, V. M. (2018). Evaluating Urban Quality: Indicators and Assessment Tools for Smart Sustainable Cities. *Sustainability*, 10(3): 1-18. https://doi.org/10.3390/su10030575.

Gomaa, A. A., & Fouad, F. M. (2022). Evaluating The Quality of Life in Urban Environments in New Cities in Egypt (Case Study: October Gardens city). *Journal of Urban Research*, 46(1): 47-7. https://dx.doi.org/10.21608/jur.2022.123228.1089.

Gülersoy, N. Z., Özsoy, A., Tezer, A., Yiğiter, R.G., & Günay, Z. (2009). Strategic Quality Planning in Historic Urban Environment. *A/Z ITU Journal of Faculty of Architecture*, 6(1): 109–125.

Hoseini, F., & Mokhtari, M. (2013). Enhancing Quality of Urban Spaces by Approach of Sense of Place (Case Study: Tarbiat Street In Tabriz City). *World Applied Sciences Journal*, 22(10):1465–1474. https://dx.doi.org/10.5829/idosi.wasj.2013.22.10.563.

Hunt, B. (2008). City Building & Placemaking. *Urban Strategies Incorporation*. www.vaughan.ca/sites/default/files/15City Building2.

Jacobs, A. B. (1993). Great Streets: Monument Avenue, Richmond, Virginia. Cambridge: MIT Press. Pages 23–27. https://escholarship.org/uc/item/3t62h1fv.

Jarvis, R. K. (1980). Urban Environments as Visual Art or As Social Settings? A Review. *The Town Planning Review*, 51(1): 50-66.

Kadhim, L. A., & Shok, M. E. (2021). Improving Social Interaction in Vertical Housing. *Iraqi Journal of Architecture and Planning*, 21(1): 135-144. https://doi.org/10.36041/iqjap.2022.174520.

Khan, A. Z., Moulaert, F., Schreurs, J., & Miciukiewicz, K. (2014). Integrative Spatial Quality: A Relational Epistemology of Space and Transdisciplinary in Urban Design and Planning. *Journal of Urban Design*, 19(4): 1–20. http://dx.doi.org/10.1080/13574809.2014.936142.

Lucas, R., & Romice, O. (2008). Representing Sensory Experience in Urban Design. *Design Principles and Practices: An International Journal*, 2(4): 83–94. http://dx.doi.org/10.18848/1833-1874/CGP/v02i04/37579.

Lynch, K. (1962). The Image of the City. *The Journal of Aesthetics and Art Criticism*, 21(1): 1–195). https://doi.org/10.2307/427643.

Marans, R. W. (2012). Quality of Urban Life Studies: An Overview and Implications for Environment-Behavior Research. *Procedia - Social and Behavioral Sciences*, 35: 9–22. https://doi.org/10.1016/j.sbspro.2012.02.058.

Nasution, A. D., & Zahrah, W. (2014). Community Perception on Public Open Space and Quality of Life in Medan, Indonesia (pp. 585–594). *Procedia - Social and Behavioral Sciences*, 153: 585-594. https://doi.org/10.1016/j.sbspro.2014.10.091.

Ministry for the Environment (2005). *New Zealand Urban Design Protocol*. https://environment.govt.nz/assets/Publications/Files/urban-design-protocol-colour.pdf.

Petroviˇ, F., & Murgaš, F. (2021). Description Relationship Between Urban Space and Quality of Urban Life. A Geographical Approach. *Land*, 10(12): 1337. https://doi.org/10.3390/land10121337.

Rouli, J., & Kusumastuti, R. D. (2020). Decreasing Urban Environment Quality and Smart Environment Initiatives. *E3S Web of Conferences* 211: 01025. https://doi.org/10.1051/e3sconf/202021101025.

Saikia, A. (2013). *Visual Dimension Summary*. Chapter 7 in the book The Dimensions of Urban Design. https://www.scribd.com/document/220070613/Visual-Dimension-Summary-by-Arnav.

Sarmento, R., Zorzal, F. M. B., Serafim, A. J., & Allmenroedr, L. B. (2005). Urban Environmental Quality Indicators. *The Sustainable City,* C. A. Brebbia A. Ferrante, M. Rodiguez & B. Terra (Editors). Pages 109–123. https://doi.org/10.2495/URS000111.

Sholihah, A. B. (2016). Overview of Urban Quality Indicators: Towards A Sustainable and Sophisticated Urban Life in Indonesia. In *Proceeding 4th International Conference on Sustainable Built and Environment* (ICSBE) (pp. 495-503). https://icsbe.uii.ac.id/4th-2016/01/downloads/archive/2016/TOPIC 3/9.

Sitte, C. (1968). City Planning According to Artistic Principles. In Camillo Sitte: The Birth of Modern City Planning, Edited by: Collins, G. R. and Collins, C. C. *The Society of Architectural Historians*, 27(1): 129–434. https://doi.org/10.2307/988439.

Talen, E. (2002). Pedestrian Access as a Measure of Urban Quality. P*lanning Practice & Research*, 17(3): 257-278. https://doi.org/10.1080/02697450220005634.

Ubaid, W. J., & AL-Ataabi, M. S. A. H. (2023). Transformations in Fashion and Architecture (Causes and Mechanisms). *The Iraqi Journal of Architecture and Planning*, 22(2): 21-39. https://doi.org/10.36041/iqjap.

Unwin, R. (1994). *Town Planning in Practice: An Introduction to the Art of Designing Cities and Suburbs*. Princeton Architectural Press. New York.

Ye, G., Yuan, H., and Zuo, J. (2021). Evaluation and Analysis of Urban Quality—A Case Study of Chongqing. In: Ye, G., Yuan, H., Zuo, J. (eds) *Proceedings of the 24th International Symposium on Advancement of Construction Management and Real Estate*. CRIOCM 2019. Springer, Singapore. http://dx.doi.org/10.1007/978-981-15-8892-1_6.

Yousif, S. S., and Al-Hinkawi, W. S. (2022). Urban Dimensions of Pedestrian Bridges in Contemporary Urban Design Ideologies. *Iraqi Journal of Architecture and Planning*, 21(1): 56-68. https://doi.org/10.36041/iqjap.2022.174282.

Chapter 2

Ventilation Systems in Multifunctional Buildings for Infectious Disease Protection

Ali K. Dawood[1,*]
and Sattar Aljabair[1]
[1] University of Technology- Iraq, Baghdad, Iraq

Abstract

In light of sustainable health care of communities, mainly after the COVID-19 pandemic, this chapter explores the vital function of ventilation systems in the field of infectious illness prevention in multipurpose structures. This focus is on how thoughtful ventilation system design and occupant placement can have a significant impact on indoor air quality, and reduction of aerosol particulate transmission and infection. The chapter analyzes how aerosol droplets from infected people spread throughout different interior environments, placing special emphasis on the significant role that properly functioning ventilation systems play. These systems strengthen defenses against infectious agents by improving general indoor air quality and reducing the risk of aerosol inhalation. Additionally, a wide range of ventilation techniques, including well-known ones like mixing ventilation, zone ventilation, stratum ventilation, and displacement ventilation, are rigorously evaluated in this chapter. These tactics are put under close examination to determine how well they work to prevent the spread of infectious diseases, with a particular emphasis on highlighting their benefits and potential drawbacks. The summarization of this research highlights the crucial function of ventilation systems and the necessity of interdisciplinary collaboration. Such partnerships are essential for

[*] Corresponding Author's Email: me.21.01@grad.uotechnology.edu.iq.

In: Advancements in Sustainability Systems
Editors: Hussain H. Al-Kayiem, Hasanain A. Abdul Wahhab and Raed A, Jessam
ISBN: 979-8-89113-571-0

improving indoor settings and guaranteeing the development of areas that prioritize public well-being. This review chapter is a priceless tool in a time when issues related to infectious diseases are still present. It is a beneficial compilation of information for policymakers, architects, engineers, and more. It emphasizes how urgently ventilation systems need to be improved, and it supports the goal of creating safer, healthier indoor settings that support public health.

Keywords: COVID-19, cough, ventilation strategies, particulate flow, Eulerian-Lagrangian modelling

1. Introduction

The world is currently grappling with the impact of a widespread respiratory virus known as H1N1 RSV, which is part of the coronavirus family that includes SARS-CoV-2. This unprecedented illness has impacted most human activities (WHO, 2020). One of the most contagious viruses ever infected humans, COVID-19 spread across most of China in just 14 days and quickly expanded worldwide in months (Wang et al., 2020). The WHO supports the droplet and contact techniques as the most common means of transmitting COVID-19. However, recent scientific studies have extensively shown the possibility of airborne transmission (Mittal et al., 2020), and the WHO does not rule out the possibility of this. The pace of expiratory events rises when a person develops a respiratory illness, boosting the production and spread of virus-carrying droplets. Talking is an example of an expiratory event besides coughing and sneezing (Morawska et al., 2009). According to the scientific data Bulfone et al., 2021, there is more inside transmission than outdoor transmission. Ventilation should be the first line of defense against infection because indoor conditions are thought to have a significant risk of spreading disease (Bhagat et al., 2020). Organizations like WHO, ASHRAE, and REHVA have guided the prevention of airborne transmission of SARS-CoV-2. These guidelines target controlling the spread of the virus through air in enclosed spaces like buildings, offices, and public transportation for the public's safety during the pandemic (ASHRAE, 2020; REHVA, 2021; WHO, 2021).

Modern research simulates the spread of COVID-19 viral droplets by using Computational Fluid Dynamics (CFD) to model an infected person's coughing (Bourouiba, 2020). For a comprehensive understanding of modeling COVID-19 transmission, it is helpful to evaluate simulations of droplet spreading through coughing and sneezing that are currently accessible. Indoor airflow, characterized by laminar, momentary, and turbulent patterns, impacts the dispersion and movement of suspended droplets (Liu et al., 2020; Liu et al., 2018; Liu et al., 2020). Additionally, human activities such as walking, coughing, and sneezing can cause fluctuations in indoor airflow. Gupta et al., (2009) conducted an experimental examination of coughing airflow dynamics. They analyzed different cough velocities over time and discovered that physiological factors such as height, weight, and gender did not significantly affect the direction of cough dispersion or the location of the lips' opening during coughing. To further reduce the risk of infection through respiratory routes, Liu et al., (2020) considered the role of human expiratory fluxes in the spread of respiratory illnesses within air-conditioned spaces. This research emphasizes the importance of accounting for human behaviors and interactions with ventilation systems in multifunctional buildings.

1.1. Indoor Air Quality and Infectious Disease Transmission

In multipurpose buildings, indoor air quality plays a key role in the spread of infectious diseases. Inadequate indoor air quality can make people more susceptible to respiratory ailments and encourage the spread of germs. Ventilation systems ought to be planned with variables like ventilation rates, filtration setups, and air exchange rates in mind (Morawska et al., 2020). A safer interior environment could be achieved by using well-maintained ventilation systems with high-efficiency filters, according to research in this area (World Health Organization, 2020).

1.2. Challenges in Multifunctional Building Environments

Buildings with many uses provide difficulties when designing ventilation systems to prevent infectious diseases. These areas accommodate a variety of uses and people with various ventilation needs. It is critical to strike a balance between indoor air quality and energy efficiency. Architectural limitations, building layouts, and flexible solutions to meet shifting occupancy patterns should also be considered when integrating ventilation systems (ASHRAE, 2020). To address these issues and provide efficient and dependable ventilation systems, a multidisciplinary approach combining architects, engineers, and healthcare professionals is necessary.

1.3. Global Response and Standards

Ventilation systems and their function in preventing disease have received more attention because of the continuing international response to the COVID-19 pandemic. Guidelines and regulations for ventilation and indoor air quality in public buildings have been produced by several organizations and regulatory agencies (CDC, 2021; WHO, 2020). With the help of these guidelines, ventilation systems could be built and maintained in a way that lowers the danger of the spread of infectious diseases. Building designers, facility managers, and lawmakers must keep up with these changing regulations since they have a significant impact on how ventilation systems are built and run in multipurpose buildings.

2. Analyses Methods

There are many investigations and analyses about air motion and virus/cough droplet transport in the air stream. Some of the methods are based on experimental investigations, and some have simulated the space and studied various parameters, like the effective distance for contamination, ventilation airflow patterns, and the effectiveness of filtration and circulation systems. A critical focus is placed on the analysis methods employed to evaluate ventilation systems in multifunctional buildings for infectious disease protection. These methods encompass comprehensive data collection, CFD simulations, and microbial analysis. By synthesizing existing research and methodologies, this review aims to offer valuable insights into the assessment of ventilation systems' efficacy in disease prevention within multifunctional building environments.

2.1. Experimental Analysis

In the realm of experimental analysis methods, this study conducted a comprehensive investigation into the dynamics of airflow and exposure to virus-laden particles in a healthcare setting. To assess the effects of different ventilation systems on the safety of healthcare workers, the researchers, as detailed by Wei et al., in 2017, meticulously examined three distinct temporal patterns of exit speed: pulsation, sinusoidal, and real cough. The experimental setup involved a range of opening diameters, spanning from 50.6 to 85.5 inches (D), simulating real-life scenarios of patients coughing.

Building upon this foundation, the study, as expounded by Kong et al., in 2021, went on to evaluate a total of sixteen distinct air distribution patterns within a typical ward scenario. These patterns were categorized into four groups: top supply and exhaust, side supply and exhaust, and bottom supply and exhaust. The critical outcome of this extensive analysis was the identification of ventilation option D, which incorporated upper diffusers for supply and lower diffusers for exhaust on the same sidewall. This particular configuration was found to be highly effective, resulting in a remarkable 70.8% reduction in cumulative exposure levels for healthcare workers.

To provide a visual reference for the experimental setup and to aid in understanding the spatial layout and arrangement of the study, Figure 1 illustrates the distances between beds and respiratory measurement points for patients in a simulated multi-patient room. This diagram serves as a valuable tool in contextualizing the study's findings, enhancing our comprehension of the experimental setup and its implications for healthcare worker safety.

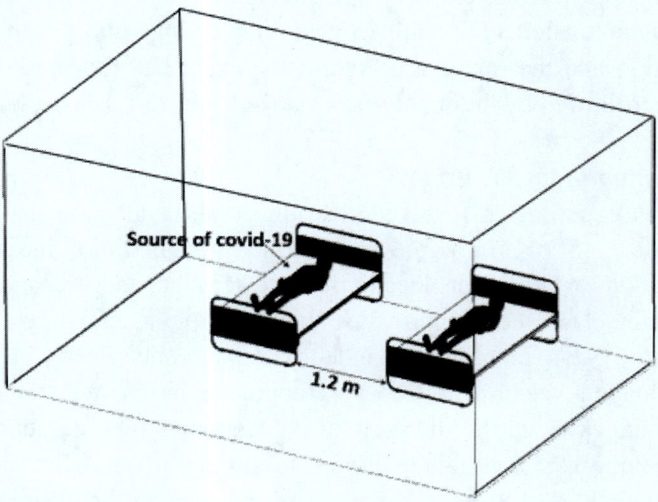

Source: authors.

Figure 1. Demonstration of the distance between beds and respiratory measurement points for patients in a multi-patient room.

2.1.1. Air Distribution

A perforated duct diffuser, a ceiling diffuser, a displacement ventilation concept using two low-velocity units, and a corridor-wall grille were the four air distribution techniques. Three test scenarios were planned for the classroom: full occupancy during the summer (heating gain of 54 W/m^2), partial occupancy during the winter (heating demand of 38 W/m^2), and full occupancy during the summer (heating gain of 40 W/m^2). The wall-grille supply idea produced high velocities (about 0.4 m/s) in the occupied zone in both summer and winter, according to the measured results of Zhao et al., 2021. For example, in airflow visualization using smoke, four air supply units are discussed in a full occupancy cooling case. Two air distribution methods rely on high momentum fluxes (corridor-wall grille and ceiling diffuser). Two strategies rely on low momentum flux (displacement ventilation unit and perforated-duct diffuser). The supply air velocities with the corridor-wall grille and ceiling diffuser were 2.9 m/s and 3.5 m/s, respectively. The initial air velocities were less than 0.2 m/s with the displacement ventilation unit and perforated-duct diffuser. Zhao et al., 2021.

Smoke flow visualization and particle image velocimetry (PIV) to examine airflow patterns. A novel laser sheet generator was used to illuminate smoke particles. The study employed the Taguchi method to evaluate the impact of supply flow rate, skirt length, and percentage of ceiling-return flow rate on recirculation. The study found that ceiling return can eliminate recirculation and that the ceiling-return flow rate had the most significant influence, followed by the supply flow rate and the skirt. Lin et al., 2020.

2.1.2. Contaminant Transport

Tracer gas experiments involve releasing a traceable gas, such as sulfur hexafluoride (SF6), into the ventilation system and measuring its concentration at different locations to determine the transport of the contaminant. This method provides insight into the effectiveness of the ventilation system in removing indoor air pollutants Zheng et al., 2018. Particle imaging velocimetry (PIV) is a popular experimental method used to measure and evaluate the transport of contaminants in ventilation systems. This technique uses laser light to illuminate particles in the airflow and capture images, which are then used to determine the velocity and direction of the airflow and the transport of contaminants (Babich et al., 2017). These experimental methods provide valuable information for understanding the transport of contaminants in ventilation systems. They can help optimize the design and performance of these systems to provide optimal protection against the spread of infectious diseases. Figure 2 illustrates the forces acting on particles as they move within the ventilation system. These forces include the influence of airflow, gravity, and drag force. By understanding how these forces interact, researchers can gain further insights into the dynamics of contaminant transport in indoor environments.

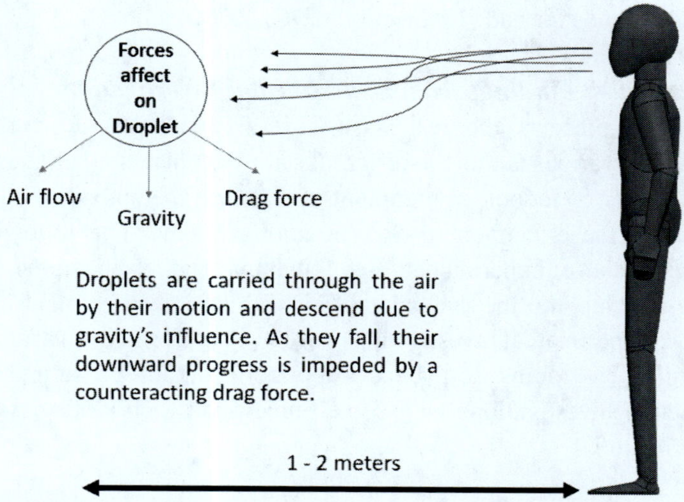

Source: authors.

Figure 2. Forces applied to the droplets during motion.

Three different temporal patterns of the exit speed (pulsation, sinusoidal, and real cough) were examined by Wei et al., 2017. The study discovered that the experimental setup's farthest reach of a real cough's flow was between a 50.6- and 85.5-inch opening diameter (D). A series of to assess the impact of various ventilation systems on the exposure of healthcare workers to virus-laden particles from patients in an enclosed chamber.

Kong et al., 2021 have experimentally evaluated 16 different air distribution patterns in a typical ward scenario and divided them into 4 categories (top supply and exhaust, side supply and exhaust, and bottom supply and exhaust). The results showed that ventilation option D, which involved upper diffusers on the sidewall supply and lower diffusers on the same sidewall return, effectively reduced cumulative exposure levels for healthcare workers by 70.8%.

2.1.3. Filtration

Aerosol challenge tests, filter efficiency measurements, and microbial and viral culture assays have been used to evaluate the efficacy of different types of filters, such as High-Efficiency Particulate Air (HEPA) filters and

Minimum Efficiency Reporting Value (MERV) filters, in capturing particles, including viruses ASHRAE, 2021. Ultraviolet C (UV-C) light is effective in inactivating viruses, including SARS-CoV-2, and can be combined with other filtration methods for added protection CDC, 2020.

2.1.4. Foremost Findings from Experimental Investigation

This section presents the results of several experimental investigations conducted by various authors in different settings. These studies shed light on crucial aspects of indoor environments, including the spread of respiratory droplets and the effectiveness of ventilation systems. The following table summarizes these experiments' key findings and applications, offering valuable insights into the dynamics of air and particle movement in various scenarios. Experimental investigations have been performed in various spaces, including office rooms, airplanes, Laboratories. Table 1 summarizes the experimental investigations with some remarks on each work based on the reported results.

Table 1. Experimental research on respiratory droplets in indoor environments

Authors	Application	Remarked
Zhu, et al., 2006	Calm indoor environment.	Results show that each cough expels more than 6.7 mg of saliva at 22 m/s and that saliva droplets can travel over 2 m.
Arumuru et al., 2020	Room	This study found that a quiet sneeze can spread over 25 feet in 22 seconds, revealing that the commonly assumed safe distance of 6 feet is not sufficient, particularly when someone sneezes.
Wang et al., 2022	Experimental Lab	Flow simulation results revealed that unsteady flow causes particle clustering. When moving upstairs, particles gather below the shoulder and quickly descend. When moving downstairs, particles disperse over the head and move far away.
Bosbach et al., 2013	Airplane Cabin	The study compared six different ventilation scenarios, including mixing ventilation, cabin displacement ventilation, ceiling-based displacement ventilation, and combined ceiling and floor-based displacement ventilation. The findings suggest that the combined operation of the floor and ceiling-based displacement ventilation system can provide high heat removal efficiencies and predictable, comfortable flow conditions.

2.2. Numerical Analysis Methods

This section underscores the pivotal role of numerical analysis methods in assessing ventilation systems within multifunctional buildings for infectious disease protection. CFD simulations serve as a primary tool, enabling researchers to visualize airflow patterns and pathogen distribution. These simulations offer insights into various ventilation scenarios, contributing to a comprehensive understanding of system performance. Additionally, numerical modeling extends to evaluating thermal comfort and energy efficiency, ensuring a holistic assessment of ventilation systems. This comprehensive approach aims to create indoor environments that prioritize safety and comfort.

A study by Al-Rawi et al., 2022 compared three scenarios where an infected person coughs: without any device and with a device placed in two different positions. The study created a contour map to illustrate the results three times during the coughing action: 0.1 seconds, 0.4 seconds, and 0.8 seconds.

2.3. Modeling of Airflow and Suspensions

Feng et al., 2020 simulation results on the social distancing effectiveness to prevent COVID-19 involving virtual humans and the effect of ambient wind on the movement of microdroplets between them. The study found that even at 3.05 meters, microdroplets can follow the airflow and settle on virtual humans' bodies and head regions when wind is present, causing recirculation. The wind speeds used in the study ranged from 0 to 16 km/h, representing different levels of wind strength, while the initial droplet diameters varied from 2 to 2000 m. The study also compared the effects of a steady-state wind and a gust with a fixed frequency. The ambient relative humidity (RH) ranged from 40 to 99.5% was also included. Additionally, the study generated a virtual person on the left with a mouth opening that had a hydraulic diameter of 10 mm and a center coordinate of 0, 0, 0.034 m.

Velocity contours of air motion in car cabins due to thermal accumulation are studied numerically by Al-Kayiem et al., 2010. They have simulated the air motion inside the car by solving the conservation governing equation using FLUENT commercial software. GAMBIT software achieved discretization of the fluid domain inside the cabin to finite volumes. Only the tetrahedral finite volume type was applied to the model. The tetrahedral discretization is a more accurate CFD simulation than the quadrilateral due to the non-uniform geometrical shape of the model under investigation. The simulation results of velocity contours are shown in Figure 3. A large air circulation cavity above the dashboard and a small air circulation cavity near the rear upper part with a stagnant portion between the two cavities were observed.

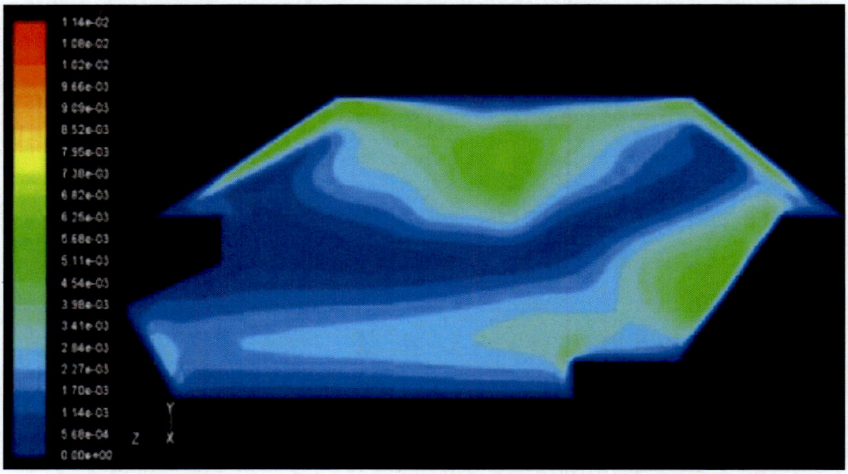

Figure 3. Velocity contours inside car cabin (with permission from Al-Kayiem et al., 2010).

Aljabair et al., 2023 performed a CFD modeling of influenza virus diffusion during coughing and breathing in a ventilated room. The simulation results are shown in Figure 4, which provides a visual representation of particle diffusion, streamlines, and temperature contours in the room during human coughing at different inlet gate angles. This visualization aids in comprehending the dynamics of cough-induced particle dispersion and temperature changes within the room, offering valuable insights for optimizing ventilation strategies and assessing infection risk.

2.4. The Eulerian-Lagrangian Modelling of Particulate Flow

The Eulerian-Lagrangian model combines two approaches to model fluid and particle behavior. The fluid phase is described using Eulerian equations, and the particle phase is modeled using Lagrangian equations (Zhu et al., 2007). The fluid phase in the model is described by the Reynolds-averaged Navier-

Stokes (RANS) equations, which simplify the mass and momentum transport equations. Reynolds stresses are usually modeled using an eddy-viscosity approach in these equations, as noted in Zhang et al., 2007.

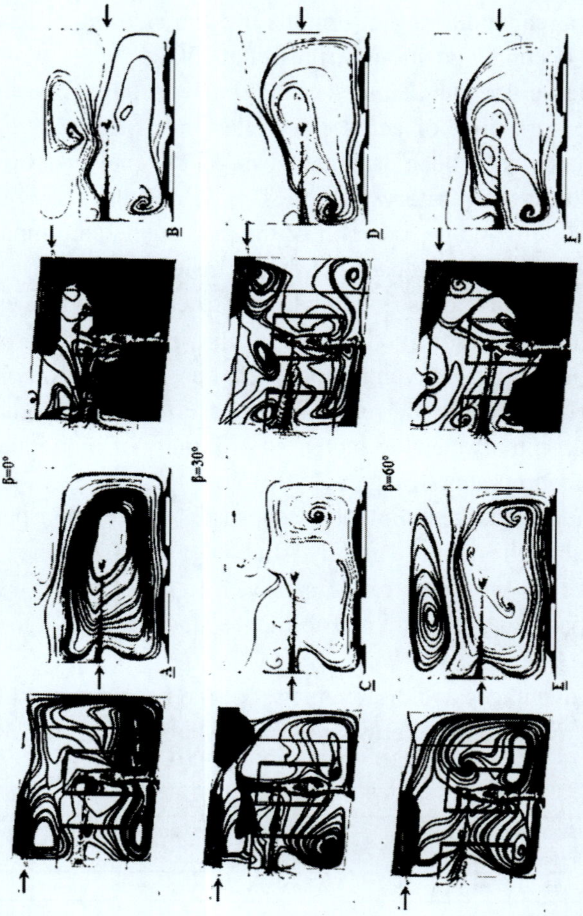

Figure 4. Particle diffusion, streamlines, and temperature contours in the room for human coughing at different inlet gate angles (Aljabair et al., 2023, with permission).

The two-phase models can be categorized into two main types, Eulerian-Lagrangian and Eulerian-Eulerian, based on how they handle the interaction between nanoparticles (solid phase) and fluid phases (Habeeb et al., 2022a, 2022b, Habeeb et al., 2023).

Eulerian-Lagrangian model to study the spread of non-uniform temperature aerosol particles in a car cabin. The model's results are compared to real-life measurements, and various turbulence models are analyzed. The validated model then simulates the spread of droplets emitted by a contagious person in a car cabin during a 15-minute trip, considering different ventilation systems and occupant positions Arpino et al., 2022.

The Eulerian-Lagrangian method has been used by Grossi et al., 2022 to simulate the spreading of aerosol particles in a passenger car cabin. The numerical model has been validated using PIV analysis to examine three scenarios during respiratory activities. The results show that the car's ventilation system and the position of the passengers can impact the amount of inhaled aerosol and the quality of the air inside the car.

The Lagrangian approach simulates the spread of droplets expelled during coughing. It is noteworthy that the results indicate uniform temperature distribution in mechanical ventilation (MV) but not in natural or displacement ventilation (DV). DV, while making it easier to expel droplets and reducing infection risk, can also lead to increased proximity of individuals, potentially elevating infection rates (Kang et al., 2015).

As a widely adopted simulation approach, Lagrangian particle tracking has been evaluated and discussed by Aljabair et al., 2021. They concluded the goodness and the weakness of the approach to simulate the particle tracking in interior space, as presented in Table 2.

Table 2. Advantages and disadvantages of the Lagrangian particle tracking Model (with permission from Aljabair et al., 2021)

Advantage	Disadvantage
To fully understand the particle's behavior and residence duration.	There are a lot of particles, which require a lot of computing.
Cheaper for a variety of particle sizes.	When turbulence is needed, the model is highly pricey.
Heat and mass transmission.	It is only allowed in small-volume fractions.
For flow, where varying particle sizes cause various particle velocities.	It is only feasible after the fact for several particles.

2.5. Foremost Numerical Investigation

Numerical studies are critical in uncovering the complexities of droplet dispersion dynamics, particularly in the context of disease transmission inside indoor environments. The researchers use complex models to study how respiratory droplets flow through the air to better understand the elements that determine infection risk. In this section, we report major findings from a series of numerical studies, mostly employing Lagrangian and Eulerian analysis methods. These studies dig into crucial aspects of droplet dispersion, such as the impact of ventilation systems, physical barriers, and individual actions on infection risk. They challenge established assumptions and provide new insights into disease transmission dynamics. Table 3 provides a concise overview of these investigations, and reference to their primary findings.

Table 3. Numerical studies on droplet dispersion dynamics and the modeling approaches

Authors	Simulation Method	Remarked
Aljabair, et al., 2021	Lagrangian	Simulations show that uncovered people in the center of a room are at higher risk of infection due to fluctuating airflow. Droplet spread is influenced by thermal conditions and the location of the AC unit, particularly if it is positioned against the coughing airflow.
Mirzaie, et al., 2021	Lagrangian	Individual seat partitions prevented infection to some extent. Increased ventilation air velocity increased droplet airflow velocities and reduced solid barrier trapping time.
Feng, et al., 2020	Lagrangian	Numerical simulations indicate that wind can complicate the spread of droplets between two virtual individuals. The simulations show that even when individuals are 3.05 meters apart, microdroplets can still flow along air currents and land on their bodies and heads.
Zhang, et al., 2019	Lagrangian	The simulation outcomes indicate that factors such as the temperature of the air supply and the relative humidity have a lesser influence on the dispersal of droplet aerosols that may carry the virus in a displacement ventilation system, compared to the rate of ventilation and the patterns in which air is distributed.
Zhang, et al., 2007	Eulerian and Lagrangian	Using the Lagrangian approach required additional processing power. When compared to the Eulerian technique, the Lagrangian one fared better.
Mariam, et al., 2021	Lagrangian	Particles from both loud and normal conversations were carried to the airflow pattern receptor. This research disproves the utility of the "rule of thumb based safe distance approach" for preventing the spread of disease.
Foster, et al., 2021	Lagrangian	The study shows that COVID-19 transmission is lower in classrooms with children than in adults due to the lower emission rates of the virus from younger children.

3. Ventilation Strategies

Ventilation strategies are an essential tool for controlling the spread of infectious diseases in multifunctional buildings. Various ventilation strategies are available, including natural ventilation, mechanical ventilation, filtration, UV-C irradiation, and combination strategies. In this section, we will discuss the advantages and limitations of each approach and provide recommendations for implementing effective ventilation strategies in multifunctional buildings.

3.1. Mixing Ventilation

Mixing ventilation (MV) design originally prioritized thermal comfort, according to ASHRAE space air diffusion publications. Most all-air conditioning systems give air 10°C below ambient temperature. Such cold air would injure a person. Hence, MV design has been to deliver this air with high velocity to maximize entrainment and mixing with the room air and to place supply diffusers much above the inhabited zone and close to the ceiling to leverage Coanda effects to minimize direct contact with the cold air jet. Fan-coil and chilled beam MV systems are all-air and air-water. MV should improve indoor air quality and temperature.

MV was most popular because the design prioritizes thermal comfort. Its design lacks energy efficiency and indoor air quality. The air handling unit must handle the space's convective heat load, which isn't always the case. Well-mixing may increase thermal comfort but not inhaled air quality, especially when there is a point source of air contaminants, as in airborne infectious disease dissemination. In summer, MV mixes room air using the supply air jet's motion, which reduces thermal comfort and energy efficiency. This analysis demonstrates that many planned advances aim to reduce MV negative effects Yang et al., 2019.

3.2. Zone Ventilation

The basic idea behind zone ventilation (ZV) is to segment the indoor space into subzones with the help of vertical jets. This can help separate contaminated areas from vulnerable ones and lower the probability of exposure to viral contaminants Cao et al., 2015. According to the investigation results by Cao et al., 2014 using ZV can reduce the risk of infection by roughly 40% compared to MV.

3.3. Stratum Ventilation (SV)

Through the inlets set into the side walls, stratum ventilation (SV) can supply air directly to the breathing zone at a higher flow rate (Tian et al., 2022). The work using simulations demonstrated that SV can dramatically lower the concentration of toxins in the zone where people breathe, successfully lowering the danger of airborne transmission (Li et al.,2020; Lu et al., 2020).

3.4. Displacement Ventilation

Displacement ventilation (DV) is an energy-saving technology that can also help the environment by replacing stale indoor air with outdoor air Tomasi et al., 2013. The results demonstrated that the DV technique was superior at clearing a room of droplets and reducing the likelihood of infection Kang et al., 2022. Figure 5 shows the difference between the Displacement and zone ventilation strategies and the direction of the airflow.

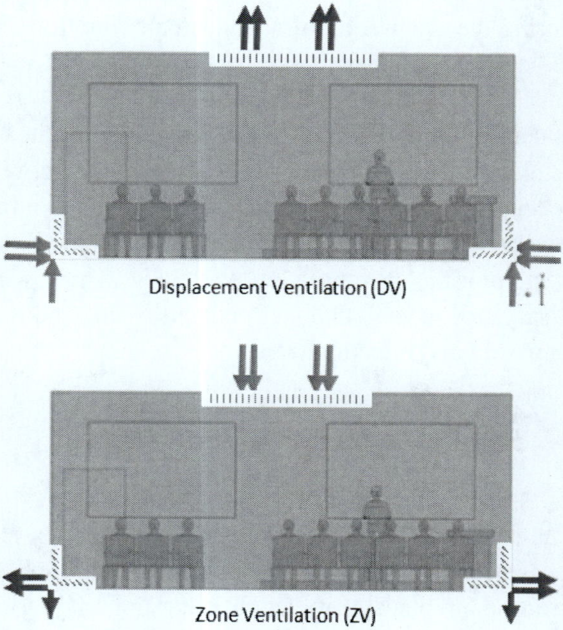

Source: authors.

Figure 5. DV and ZV strategies of ventilation.

4. Author Suggestions

I. Further Validation and Refinement of the Model: Future research should focus on validating and refining the Eulerian-Lagrangian model by comparing it with additional real-life measurements and experimental data.

II. Investigation of Specific Infectious Agents: Future work should explore the transmission dynamics of specific infectious agents by considering their unique characteristics and behavior to design targeted ventilation strategies.

III. Consideration of External Environmental Factors: The influence of outdoor air quality, wind speed, and temperature on particle dispersion should be incorporated into the modeling approach.

IV. Experimental Validation of Model Parameters: Conducting experiments to validate and refine the model parameters used in the Eulerian-Lagrangian model would enhance its accuracy and applicability.

V. Integration of Advanced Filtration and Disinfection Technologies: Exploring the integration of advanced air filtration and disinfection technologies within ventilation systems can lead to more efficient approaches for mitigating airborne transmission.

VI. Consideration of Human Behavior and Occupant Dynamics: Investigating the influence of human behavior, movement patterns, and occupancy density on particle dispersion can inform the design of ventilation systems. Long-term Monitoring and Evaluation: Long-term monitoring of ventilation strategies in real-world settings is necessary to assess their effectiveness in controlling airborne transmission and infection rates.

Conclusion

This chapter underscores the pivotal role of ventilation systems in indoor infectious disease control. It emphasizes the need for careful system design and optimal occupant placement. Our exploration of various ventilation strategies highlights their pros and cons, stressing the importance of interdisciplinary collaboration among experts in architecture, engineering, and healthcare.

Looking ahead, the paper points to critical research areas, including model validation, consideration of specific pathogens, external factors like outdoor conditions, experimental parameter validation, and integration of advanced filtration and disinfection technologies. This review ultimately emphasizes that ventilation systems are central to safeguarding public health in multifunctional buildings, contributing to a safer indoor environment for all.

References

Alesbe, I., Abdul Wahhab, H. A., & Aljabair, S. (2023). Transient Study of Thermal Stratification of Full-Scale Chilled Water Storage Tank During Optimum Discharge Condition. *Journal of Energy Storage*, 65: 107236. https://doi.org/10.1016/j.est.2023.107236.

Al-Kayiem, H. H., Sidik, M., & Munusamy, Y. (2010). Study on the Thermal Accumulation and Distribution Inside a Parked Car Cabin. *American Journal of Applied Sciences*, 7(6). https://doi.org/10.3844/ajassp.2010.784.789.

Al-Rawi, M., Al-Jumaily, A.M., & Lazonby, A. (2022). Did You Just Cough? Visualization of Vapor Diffusion in an Office Using Computational Fluid Dynamics Analysis. *International Journal of Environmental Research and Public Health*, 19(16): 9928. https://doi.org/10.3390/ijerph19169928.

Aljabaır, S., Alesbe, I. & Alkhalaf, A. (2023). CFD Modeling of Influenza Virus Diffusion During Coughing And Breathing in a Ventilated Room. *Journal of Thermal Engineering*, 9 (1): 127-137. https://doi.org/10.18186/thermal.1243491.

Arpino, F., Cortellessa, G., Grossi, G., & Nagano, H. (2022). A Eulerian-Lagrangian Approach for the Non-Isothermal and Transient CFD Analysis of The Aerosol Airborne Dispersion In A Car Cabin. *Building and Environment*, 209: 108648. https://doi.org/10.1016/j.buildenv.2021.108648.

Arumuru, V., Pasa, J., and Samantaray, S. S. (2020). Experimental Visualization of Sneezing And Efficacy of Face Masks And Shields. *Physics of Fluids*, 32(11): 115129. https://doi.org/10.1063/5.0030101.

ASHRAE. (2020). ASHRAE Position Document on Infectious Aerosols. Position Documents. Retrieved from the ASHRAE website: Position Documents (ashrae.org).

ASHRAE. (2021). ASHRAE Position Document on Infectious Aerosols. Position Documents. Retrieved from the ASHRAE website: Position Documents (ashrae.org).

Bhagat, R., Davies Wykes, M., Dalziel, S., and Linden, P. (2020). Effects of Ventilation on the Indoor Spread of COVID-19. *Journal of Fluid Mechanics*, 903, F1. https://doi.org/10.1017/jfm.2020.720.

Bourouiba, L. (2020). Turbulent Gas Clouds and Respiratory Pathogen Emissions: Potential Implications for Reducing Transmission of COVID-19. *JAMA*, 323(18): 1837-1838. https://doi.org/10.1001/jama.2020.4756.

Bulfone, T. C., Malekinejad, M., Rutherford, G.W., and Razani, N. (2020). Outdoor Transmission of SARS-CoV-2 and Other Respiratory Viruses: A Systematic Review.

The Journal of Infectious Diseases, 223(4): 550-561. https://doi.org/10.1093/infdis/jiaa742.

Cao, G., Nielsen, P. V., Jensen, R. L., Heiselberg, P., Liu, L., and Heikkinen, J. (2015). Protected Zone Ventilation And Reduced Personal Exposure to Airborne Cross-Infection. *Indoor Air*, 25(3): 307-319. https://doi.org/10.1111/ina.12142.

Cao, G., Sirén, K., and Kilpeläinen, S. (2014). Modelling and Experimental Study of Performance of the Protected Occupied Zone Ventilation. *Energy and Buildings*, 68, 515-531. https://doi.org/10.1016/j.enbuild.2013.10.008.

Centers for Disease Control and Prevention (CDC). (2021). *Ventilation in Buildings.* Retrieved from https://www.cdc.gov/coronavirus/2019-ncov/community/ventilation.html.

Feng, Y., Marchal, T., Sperry, T., & Yi, H. (2020). Influence of Wind And Relative Humidity on the Social Distancing Effectiveness to Prevent COVID-19 Airborne Transmission: A Numerical Study. *Journal of Aerosol Science*, 147: 105585. https://doi.org/10.1016/j.jaerosci.2020.105585.

Foster, A., and Kinzel, M. (2021). SARS-CoV-2 Transmission In Classroom Settings: Effects Of Mitigation, Age, And Delta Variant. *Physics of Fluids*, 33(11): 113311. https://doi.org/10.1063/5.0067798.

Gupta, J. K., Lin, C. H., and Chen, Q. (2009). Flow Dynamics And Characterization of a Cough. *Indoor Air*, 19(6): 517-525. https://doi.org/10.1111/j.1600-0668.2009.00619.x.

Hamza, N F. A., and Aljabair, S. (2022). Evaluation of Thermal Performance Factor By Hybrid Nanofluid And Twisted Tape Inserts In Heat Exchanger. *Heliyon*, 8(12). https://doi.org/10.1016/j.heliyon.2022.e11950.

Hamza, N. F. A., and Aljabair, S. (2023). Numerical and Experimental Investigation of Heat Transfer Enhancement by Hybrid Nanofluid and Twisted Tape. *Engineering and Technology Journal*, 41(1): 69-85. https://doi.org/10.30684/etj.2022.131909.1069.

Kang, Z., Zhang, Y., Fan, H., and Feng, G. (2015). Numerical Simulation of Coughed Droplets in the Air-Conditioning Room. *Procedia Engineering*, 121:114-121. https://doi.org/10.1016/j.proeng.2015.08.1031.

Kong, X., Guo, C., Lin, Z., Duan, S., He, J., Ren, Y., and Ren, J. (2021). Experimental Study On The Control Effect of Different Ventilation Systems on Fine Particles in a Simulated Hospital Ward. *Sustainable Cities and Society*, 73: 103102. https://doi.org/10.1016/j.scs.2021.103102.

Li, H., Xi, C., Kong, X. Lin, Z. and Wang, L. (2020). A Comparative Experimental Investigation On Radiant Floor Heating System And Stratum Ventilation. *Sustainable Cities and Society*, 52: 101823. https://doi.org/10.1016/j.scs.2019.101823.

Lin, T., Zargar, O. A., Lin, K-Y. Juiña, O., Sabusap, D. L., Hu, S-C. and Leggett, G. (2020). An Experimental Study of The Flow Characteristics And Velocity Fields In An Operating Room with Laminar Airflow Ventilation. *Journal of Building Engineering*, 29: 101184. https://doi.org/10.1016/j.jobe.2020.101184.

Liu, F., Qian, H., Luo, Z., Wang, S. and Zheng, X. (2020). A laboratory Study of The Expiratory Airflow And Particle Dispersion In The Stratified Indoor Environment. *Building and Environment*, 180: 106988. https://doi.org/10.1016/j.buildenv.2020.106988.

Liu, W., and Chen, Q. (2018). Development of Adaptive Coarse Grid Generation Methods For Fast Fluid Dynamics In Simulating Indoor Airflow. *Journal of Building Performance Simulation*, 11(4): 470-484. https://doi.org/10.1080/19401493.2017.1397195.

Liu, W., Hooff, T. V., An, Y., Hu, S., and Chen, C. (2020). Modeling Transient Particle Transport In Transient Indoor Airflow By Fast Fluid Dynamics With The Markov Chain Method. *Building and Environment*, 186; 107323. https://doi.org/10.1016/j.buildenv.2020.107323.

Magar, A. M, Joshi, M., Rajagopal, P. S., Khan, A., Rao, M. M., and Sapra, B. K. (2021). CFD Simulation of the Airborne Transmission of COVID-19 Vectors Emitted during Respiratory Mechanisms: Revisiting the Concept of Safe Distance. *ACS Omega*, 6(26): 16876-16889. https://doi.org/10.1021/acsomega.1c01489.

Mittal, R., Ni, R., and Seo, J. H. (2020). The flow physics of COVID-19. *Journal of Fluid Mechanics*, 894(F2). https://doi.org/10.1017/jfm.2020.330.

Morawska, L., Johnson, G. R., Ristovski, Z. D., Hargreaves, M., Mengersen, K., Corbett, S., Chao, C. Y. H., Li, Y., and Katoshevski, D. (2009). Size Distribution and Sites of Origin of Droplets Expelled From The Human Respiratory Tract During Expiratory Activities. *Journal of Aerosol Science*, 40(3): 256-269. https://doi.org/10.1016/j.jaerosci.2008.11.002.

Morawska, L., Tang, J. W., Bahnfleth, W., Bluyssen, P. M. (2020). How Can Airborne Transmission Of COVID-19 Indoors Be Minimised?. *Environment International*, 142: 105832. https://doi.org/10.1016/j.envint.2020.105832.

REHVA. (2021). *REHVA COVID-19 Guidance Version 4.1.* Retrieved from https://www.rehva.eu/activities/covid-19-guidance/rehva-covid-19-guidance.

Tian, X., and Lin, Z. (2022). Dynamic Modeling Of Air Temperature In The Breathing Zone with Stratum Ventilation Using a Pulsating Air Supply. *Building and Environment*, 210: 108697. https://doi.org/10.1016/j.buildenv.2021.108697.

Tomasi, R., Krajčík, M., Simone, A., and Olesen, B. W. (2013). Experimental Evaluation of Air Distribution In Mechanically Ventilated Residential Rooms: Thermal Comfort And Ventilation Effectiveness. *Energy and Buildings*, 60: 28-37. https://doi.org/10.1016/j.enbuild.2013.01.003.

Wang, D., Zhou, M., Nie, X., Feng, X., Xiao, Y., and Chen, W. (2020). Epidemiological Characteristics and Transmission Model of Coronavirus Disease 2019 in China. *Journal of Infection*, 80(5): 25-27. https://doi.org/10.1016/j.jinf.2020.03.008.

Wang, H., Li, Z., Liu, Y., Zhu, L., and Zhou, Z. (2022). Experimental Study of the Dispersion of Cough-Generated Droplets from a Person Going Up- or Downstairs. *AIP Advances*, 12(1). https://doi.org/10.1063/5.0073880.

Wei, J., and Li, Y. (2017). Human Cough as a Two-Stage Jet and Its Role in Particle Transport. *PLOS ONE*, 12(1), e0169235. https://doi.org/10.1371/journal.pone.0169235.

WHO. (2021). *Roadmap to Improve and Ensure Good Indoor Ventilation in the Context of COVID-19*. Retrieved from https://apps.who.int/iris/handle/10665/339857.

World Health Organization (WHO). (2020). *Guidelines on the Use of Masks In The Context of COVID-19.* Retrieved from https://www.who.int/emergencies/disease-outbreak-news/item/2020-DON253.

World Health Organization. (2020). *Transmission of SARS-CoV-2: Implications for Infection Prevention Precautions.* Retrieved from https://www.who.int/news-room/commentaries/detail/transmission-of-sars-cov-2-implications-for-infection-prevention-precautions.

World Health Organization. (2020). *Coronavirus Disease (COVID-2019) Situation Reports.* Retrieved from: https://www.who.int/emergencies/disease/novel-coronavirus-2019/situation-reports/.

Zhao, W. Lestinen, S., Mustakallio, P., Kilpeläinen, K., Juha Jokisalo, J., and Kosonen, R. (2021). Experimental Study on Thermal Environment in A Simulated Classroom with Different Air Distribution Methods. *Journal of Building Engineering*, 43: 103025. https://doi.org/10.1016/j.jobe.2021.103025.

Zhang, Y., Feng, G., Bi, Y., Cai, Y., Zhang, Z., and Cao, G. (2019). Distribution of Droplet Aerosols Generated by Mouth Coughing and Nose Breathing in an Air-Conditioned Room. *Sustainable Cities and Society*, 51: 101721. https://doi.org/10.1016/j.scs.2019.101721.

Zhang, Z., and Chen, Q. (2007). Comparison of the Eulerian and Lagrangian Methods for Predicting Particle Transport in Enclosed Spaces. *Atmospheric Environment*, 41: 5236-5248. https://doi.org/10.1016/j.atmosenv.2006.05.086.

Zhang, Z., Zhang, W., Zhai, Z. J., & Chen, Q. Y. (2007). Evaluation of Various Turbulence Models in Predicting Airflow and Turbulence in Enclosed Environments by CFD: Part 2—Comparison with Experimental Data from Literature. *HVAC&R Research*, 13(6), 871-886. https://doi.org/10.1080/10789669.2007.10391460.

Zhu, H. P., Zhou, Z. Y., Yang, R. Y., & Yu, A. B. (2007). Discrete Particle Simulation of Particulate Systems: Theoretical Developments. *Chemical Engineering Science*, 62(13): 3378-3396. https://doi.org/10.1016/j.ces.2006.12.089.

Zhu, S., Kato, S., and Yang, J.-H. (2006). Study on Transport Characteristics of Saliva Droplets Produced By Coughing in a Calm Indoor Environment. *Building and Environment*, 41(12): 1691-1702. https://doi.org/10.1016/j.buildenv.2005.06.024.

Chapter 3

Impact of the Generative Design Approach on the Sustainability of Urban Patterns

Afnan R. Mohsin[1,*]
Khansaa G. Rasheed[1]
and Zuhair A. Nasar[2]
[1]Department of Architectural Engineering, Engineering Faculty,
University of Technology- Iraq, Baghdad, Iraq.
[2]Faculty of Physical Planning, University of Kufa, Kufa, Iraq.

Abstract

Generative design is a contemporary approach urban areatilizes algorithms to automate the design process for a sustainable urban. It allows designers to test a wide range of alternatives and manage numerous variables and parameters that are difficult to manage manually. The traditional routine approach has a gap between practice and contemporary requirements, causing it to be inflexible for addressing complex urban problems, necessitating new design methods and tools to determine the relationships between previous practices and the current and secure sustainability. This chapter is aimed at qualitative measurements in both collecting and analyzing data. Data have been collected from the literature to study the shifting in approaches down to the contemporary methods in urban design, followed by the analysis of Kartal Sub-Center and Kartal-Pendik Waterfront that adopted the generative approach. Then, the transitions between the different design stages and the final output are identified to determine the sequences of the generative process. The results demonstrate that generative design

---

[*] Corresponding author email: ae.20.38@grad.uotechnology.edu.iq.

In: Advancements in Sustainability Systems
Editors: Hussain H. Al-Kayiem, Hasanain A. Abdul Wahhab and Raed A, Jessam
ISBN: 979-8-89113-571-0

effectively generates urban patterns at all stages, moving from deterministic to probabilistic in the urban pattern and integrating the up-down and bottom-up concepts into a unified conceptual framework. From this review, the chapter concludes with a new methodology depending on the role of the generative design approach in creating sustainable, flexible, and livable urban cities.

Keywords: generative system, urban pattern, parametric design, optimizing level, generation level

1. Introduction

Generative design was first proposed in the 1970s and used in architectural design in 1974 (Frazer, 2002). Its methods help designers create new designs automatically based on rules. In recent years, the techniques have been introduced in many applied fields, including engineering fields, which, in turn, has reduced routine work. At the same time, traditional approaches are often limited in the specialization of the problem. The design of future cities is one of the most significant challenges, and the use of advanced technologies, such as generative design tools, can help us move from a traditional policy-making framework to an analytical process with the potential for collaboration between different stakeholders. Generative design capabilities allow data analysis and interpretation, the generation of complex shapes, and automatic evolution processes. In addition, it is based on various methods, systems, and theories. Technologies are also influenced by multiple sciences, such as biology, algorithms, and engineering, and it is not related to the preparation of a plan but rather to the establishment of the process or mechanism that leads to the generation of the output (İnceoğlu et al., 2022).

A variety of computational methods have been proposed at different time stages, including different forms of spatial analysis to evaluate the performance of design proposals, in addition to environmental analysis and simulation of building behavior or the automated generation of urban design proposals based on specific criteria, with the help of a different set of programs such as Rhino & Grasshopper 3D, City Engine. However, the challenge of integrating urban analysis and generative methods is to create an appropriate computational representation of an urban design problem. Algorithmic systems serve as the foundation for all generative systems and describe the sequence of operations required to execute specific tasks. However, there is frequently room for improvement in understanding these two ideas (algorithms systems, computation). Computing may be seen as a procedural act for tasks, different from algorithms in that it doesn't require the usage of computers (Fernandes, 2013). Nevertheless, computers are a tool for engineers to overcome time and diversity limitations in design solutions. However, there remain limitations of optimization algorithms, grammar, and performance (Han et al., 2021).

By reviewing the previously available literature, it has been explored that the gap in current practices in urban design because of the continuous changing of contemporary life requirements. So, the new approach has been analyzed step-by-step, starting from the theoretical base and secondary approaches. Then, the stages of it explore the role of generative systems in urban design.

2. Urban Pattern

In the 20th century, urban design was based on the development of manual control. Pre-computer technology has contributed to the development of documentation in urban planning. Automated design systems (and later geographic information systems) set strict requirements for models of "data structuring, classification of basic spatial design processes" and urban environments. At the same time, theories focused on iterations and replicating physical elements and prototype processes (Gashenko, 2019).

2.1. Traditional Urban Pattern

In general, Urban design is a specialization that involves the interaction of components and elements to influence the values and meanings of urban products and their perception (Najm & Al-Yousif, 2021). In their book, Alexander et al., 1977 presented more than 250 models for planning and designing urban spaces, including a set of typical problems and solutions. His main idea was based on generalizing the common phenomenon with accuracy in its classification and providing a means of solving it through the presence of elements used instead of searching for new solutions. Types of urban patterns are summarized in Table 1 and discussed in detail in the following subsections.

Table 1. The basic urban patterns

Type of pattern	Description	Pattern outline and shape
Nukleus	a point type: The nucleus pattern is the presence of a strong and independent building. It is assumed that the inner courtyard meets this function for clustered buildings. A nucleus often indicates the start of subsequent settlement activity.	
Cluster	a stochastic field type: With this type of pattern, it is considered that Site development, parceling, and buildings are not fixed from the beginning, as they are subject to steps and development processes	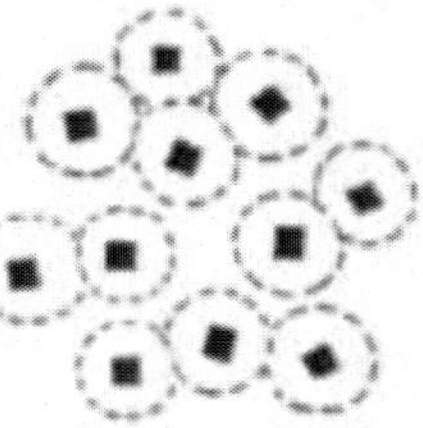
Wegelagerer	A passive linear type: - A linear building structure is formed by successive additions, such as houses next to each other.	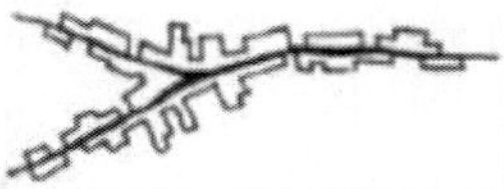
Ausleger	An active linear type: - This pattern can be used in existing areas, old paths, or on one large plot of land.	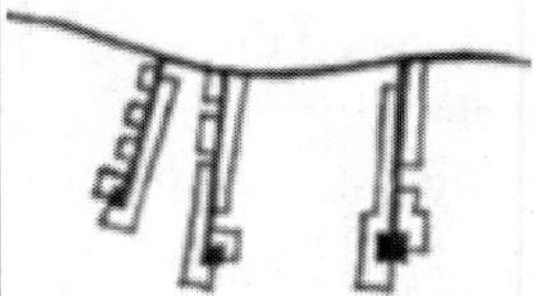
Vernetzer	An interactive field type: - The urban pattern is created based on old paths or overland roads, and some road junctions, "cliffs," and winding roads are gradually being eliminated.	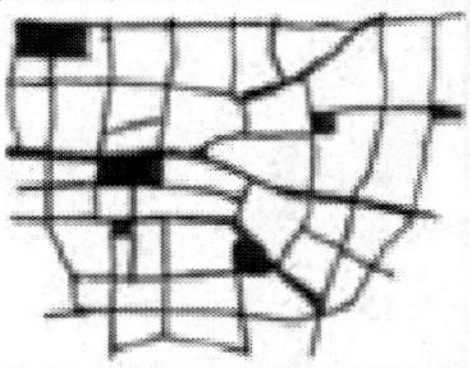
Plan	A deterministic field type: - is characterized by a plan that defines the division, structure, and development of the site, as well as the architectural design for each building.	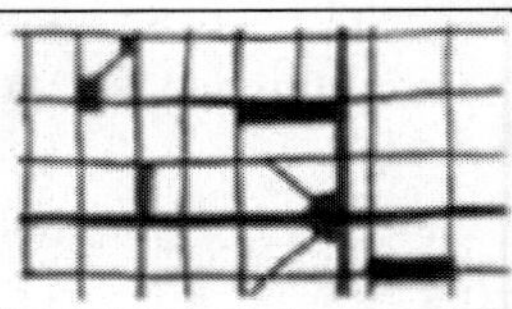

2.1.1. Basic Types of Urban Patterns

The basic urban patterns came from the concept of *Archetype*, which is defined as a typical example of something or the original model of something from which others are copied, as outlined by the Cambridge Dictionary. In his book

"Town Square," Zucker 1959 first systematically used Archetypes in architectural theory. The prototyping theory was further developed in the sixties in Aldo Rossi's book "The Architecture of the City" in 1966. Moreover, the primary traditional urban patterns (which function as indicators of urban patterns) depend to some extent on fixed forms, representing a snapshot of a form at any moment. They were viewed as a visually displayed design manual that combined the types of permissible roadways and the connections between them. They are also included within the characteristics of urban elements, such as the type of streets, the intersection of land uses, or the treatment of urban design (Marshall, 2005). Traditional cities did not invent new urban patterns to distinguish them from others. Still, their distinction was based on how they dealt with what existed, so they adhered to their proportions and characteristics (Koenig, 2018), as shown in Table 1.

2.1.2. Shifting Approaches in Urban Design

Returning to the traditional approaches in architecture and urban design, we note that since the 18th century, it has been deriving its ideas from returning to the origins of natural architecture and then entering it into different stages up to its emergence because of the industrial era. Based on (architecture as a process for producing its functional parts), urban designers consider that urban pattern is an inert future as there is a system imposed on it from the outside. Therefore, urban style results from fixed models (Stavric & Marina, 2012). Due to rapid development, digital design techniques have expanded both technical and analytical capacity within practices and theories in urban design. It has also allowed the exploration of complex architectural systems and the introduction of effective, innovative design possibilities. The generative design approach deals with many ideas, approaches, conceptual frameworks, and logical processes that lead to other applications that provide data, models, and visualizations, not just approaches with specific steps and outputs (Boumaraf, 2022). Jane Jacob, in 1961, in the last chapter of her book, expresses the period in which architecture and urban design began to pursue another path in productivity as a period that resulted from the interaction of an infinite number of generated variables. Going back to the nature of the pattern in different periods, for example, in classicism (De Quiney), the pattern was considered the perception that carries thought, which is considered as a basis for establishing the form and that it remains constant without change in all the models resulting from it. In modernity, it was considered the ideal form for specific (inevitable) use (Salman and Rasheed, 2018). In postmodernism, the pattern was considered an essential but non-defining factor for the established form (Al-Kazzaz, 1999). In generative approaches, with dynamic analysis, they transform design problems into a continuous process with an open end, thus shifting from the inevitable in making the design decision to the non-inevitable (Ayaroglu, 2007). In addition to this transition, there has been a change in the traditional view of cities from static and centralized to unstable and complex, in addition to being more dynamic and flexible. Thus, the shift from the fixed specific image to the undefined images through dynamic processes. However, both approaches share common concepts, such as self-organization.

3. Generative Design

The generative system in urban design began in 1987 as a small volume by Alexander et al., considered a landmark in the design patterns movement called this theory "new theory." The generative design method focuses on gradually creating a pattern through various disciplines. The design was considered a gradual, non-linear development in response to contextual urban factors. Alexander's ideas were combined and challenged, leading to the development of the generative design. Generative design is an iterative process in which the designer identifies the computational model and its design goals; the computer generates various design alternatives. Parameters, one of the generative components, help designers understand the relationship between parameters and goals, leading to better design solutions (Gradišar et al., 2022).

The generative design consists of two main components: parametric model and optimization. Generative design simulation integrates both indicators to generate near-perfect solutions for strategically placing facilities and urban spaces (Kumalasari et al., 2022). Another definition of generative design is the process that enables users to apply various restrictions and inputs to a specific system without supplying a full technical model. This generative model offers its architecture and replicates designs while adhering to the instructions provided. The process produces design materials such as 3D models, graphics, and simulations, as shown in Figure 1 (Westerveld, 2021).

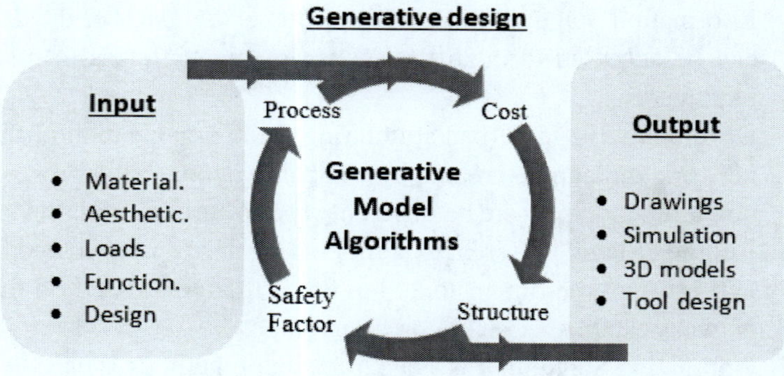

Figure 1. The generative design process (reproduced by authors with permission from Westerveld, 2021).

3.1. Stages to Generate Urban Patterns

Despite the multiplicity or diversity of the tools in generative design and thus the diversity of the product, it generally consists of four basic stages, as laid out in Figure 2.

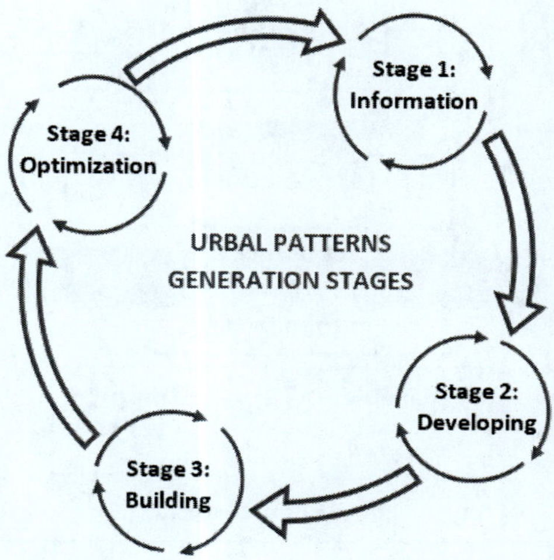

Figure 2. The urban pattern generation stages.

As deliberated by Koenig & Bauriedel, 2004, secondary branches of tools appear in certain tools over others. The stages for urban pattern development are the following:

I. Information Stage: In this stage, the contextual basis of the data bank can be accessed through the system of higher levels to modify its parameters.
II. Developing Stage: Urban structuring is the organization of the street network and can be created by controlling connection processes and combining them to create different types of urban structures.
III. Building Stage: Different experiments are used and developed around generating three-dimensional building structures based on the street network established in the previous stage.
IV. Optimizing Stage: In this stage, a group of methods is used to rebuild structures according to special criteria, such as land use, solar radiation, ventilation, land divisions, and road networks.

Generally, the nature of the work of generative design programs consists of basic stages. These stages start with the input and end with the final output, and between them are the approved processes (Abussaud & Alshuwaikhat, 2023), as in Figure 3. explained that the generative design process consists of basic stages:

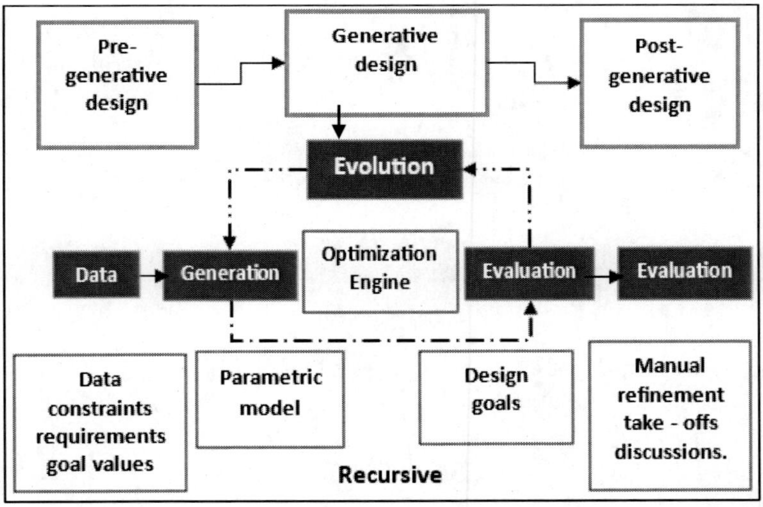

Figure 3. The three stages of the generative design framework, authors adapted from (Abussaud & Alshuwaikhat, 2023).

A. Generation - where a specific design space is identified as a closed system that can generate all possible solutions to a specific design problem.
B. Evaluation — in which metrics are developed to judge and evaluate the performance of each design.
C. Evolution — At this stage, evolutionary algorithms search the design space between the generated solutions to find high-performing designs.

3.2. Mathematical Theories of the Generative System

The most important of these theories Began with the theory of Euclidean geometry and Gaussian theory in determining the curvature of surfaces and the theory of the fourth dimension of geometry, the following as the Mathematical theories (Orabi, 2016):

A. Simulation is "the process of designing a model that resembles another original element or system, then experimenting with it to fully understand the behavior of the original system or element, and then evaluating different operating strategies on that system." Simulation is a set of techniques or tools that use computers to simulate natural processes. A system with rules and assumptions about its performance must be developed. These assumptions are used to experiment with the behavior of the building or model.
B. Generation: automatically creating a specific shape or idea using the generative or algorithmic system. Programs are used to support the volume development process.
C. Design Algorithms: Generally, an algorithm contains a series of steps written in a basic coding language and controlled by precise instructions that move in different steps (1-2-3). Furthermore, the algorithm is a sequential set of logical thinking processes used to make design decisions governed by laws and standards.
D. Parameters: Oxford Dictionary defines them as "measurable numerical or non-numerical parameters that form part of a set of elements that define a particular system or a set of working conditions." In computer definition, it is a variable that must be set to a specific value to start executing a program. In digital design,

parameters are the building blocks of algorithms, and the best algorithms have better definitions of determinants and variables.

E. Variables are any property, factor, or element subject to change or modification computationally and considered as data that takes one value or more during one or more periods of the program.
F. Performance Evaluation is the process of evaluating the success of a design model or part of it against the mentioned requirements, which were reached during the analysis phase. The evaluation process can be performed after implementation or after use to evaluate the functional characteristics of the model to verify the quality of the design and implementation.
G. Cybernetics Norbert Wiener defines cybernetics as "the science of communications and automatic control systems of machines and animals."
H. Simplexity: A system results from the superposition of a complex set of rules. It has recently been found to be used by algorithms and text designers to create complex shapes from a series of simple mathematical symbols and equations.
I. Continuous Surfaces: occur when merging three elements (walls, columns, and floors) leads to continuous surfaces. One of the first examples of these surfaces is (Endless House), as in Figure 4a, designed by architect (Friedric Kiesler).
J. Mesh Surface: It is a digital method for creating complex polygonal or polyhedral shapes. It depends on several aspects, including the arrangement of vertices, edges, and faces, to determine the desired output, as shown in Figure 4b.

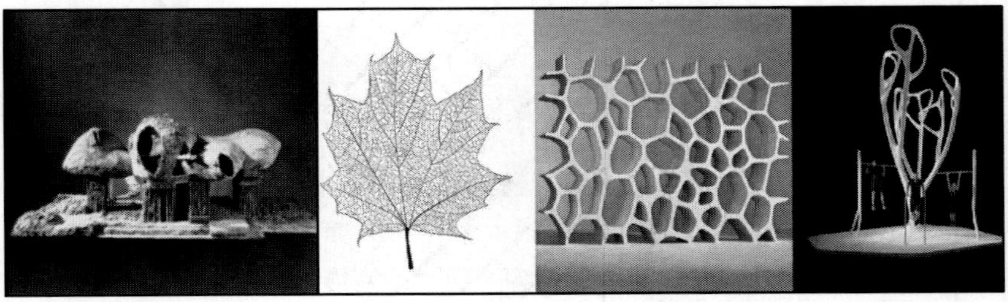

a b

Figure 4. Mathematical theories. a. endless house design, b. mesh surface (Orabi, 2016).

K. NURBS: representing the strong curvature model and as a mathematical model for representing and generating curves and surfaces. Also, it can be controlled by checking their control points.
L. Morphogenesis: The origin of the word Morphogenesis is derived from the Latin origin (morph) and (genesis), which means formation or form, and in the literal sense, means the beginning of the formation or birth of a form. As for scientific significance, it means the processes that modify the shape of the living organism.

Terzdis, 2006 explained that it is the process in which the transformation of an element from one form to another occurs gradually. The goal of this process is not the appearance of the final form but rather the process itself. (Jović & Mitić, 2020).

3.3. Approaches of Generative System in Urban Design

Generative design has emerged as a tool for urban design, with steps that proceed according to a specific approach. One of the research attempts to classify curricula for generative urban design methodologies showed that there is a set of approaches (El-Zanfaly, 2009), which are:

A. Deforming a basic shape. According to this approach, reliance is generally on generative parametric design. The designer proposes an initial shape modified parametrically to produce a final alternative to the design, as pointed out by Beirão & Duarte, 2005.
B. Simulation of real urban growth. This approach to generative design methodologies attempts to produce algorithms that simulate the engineering logic of an urban model's emergence and growth processes. A sample is illustrated in Figure 5.
C. Urban form as building blocks. This approach investigates the possibility of creating 3D urban forms based on applying basic rules, as shown in Figure 6.
D. Analyzing a realistic spatial structure. According to this approach, the generative methodology starts from the results of an analytical process of the spatial structure of the urban pattern, as it proposes changes in the spatial structure of the urban pattern using what is related to the numerical results of the analysis, as shown in Figure 7.

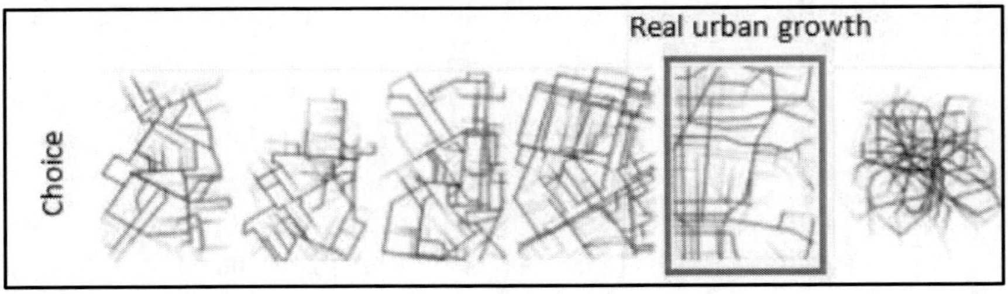

Figure 5. Simulation of real urban growth (Shehu, 2017).

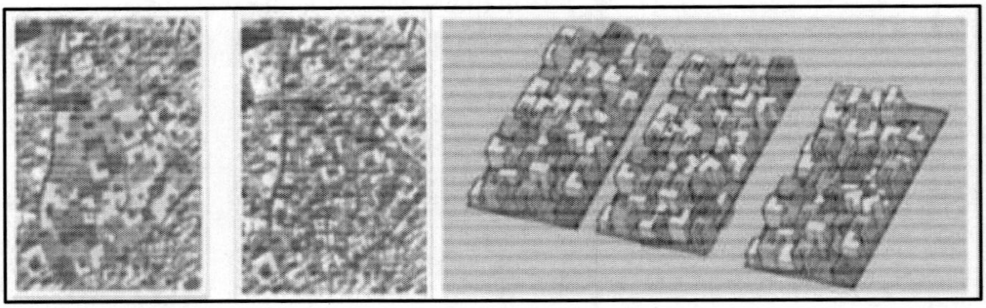

Figure 6. Urban form as building blocks (Al-Sayed, 2013). Synthetic Space Syntax: A generative and supervised learning approach in urban design.

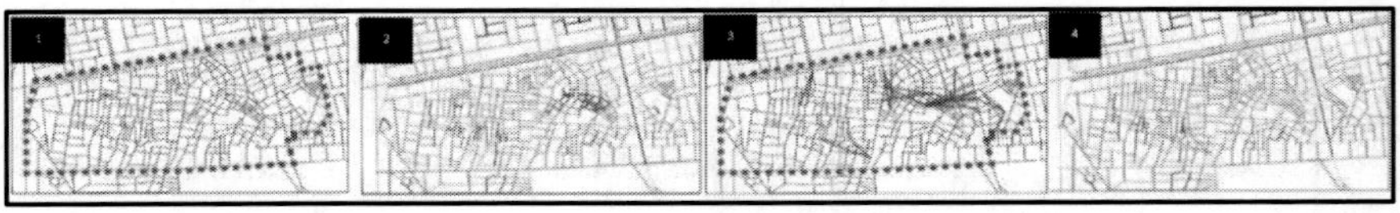

Figure 7. Analyzing a realistic spatial structure (Abdulqader et al.,2009). 1) initial axial lines map, 2) R5 integration analysis of the initial axial map, 3) highlighted new axial lines for improvement, 4) R5 integration analysis of improved axial line map.

In addition, the existence of other approaches based on previous theories, such as (Grammars for designing), was built on (shape grammars), and that formal generative system for producing design languages (Qeisi & Al-Alwan, 2021). This system was introduced by Stiny and Gips, 1972. Shape Rules are a set of rules for shape transformation in addition to being a generative engine that selects and processes these rules step by step, starting from the initial shape to creating a design language, as shown in Figures 5 to 8.

Figure 8. Grammar for designing (based on Qeisi & Al-Alwan, 2021).

This approach has been used to represent an approach to design using the grammatical rules of form not derived from current urban design. In some cases, the analysis of existing urban plans is not used to define grammatical rules but to define a general design methodology. Some of its directions move towards the concept of flexible urban design, where flexible urban design involves creating a complete design system that can be modified and adapted to various contexts (Mandić and Tepavčević, 2015).

4. Case Study: Kartal Sub-Center and Kartal-Pendik Waterfront

Kartal Sub-Center and Kartal-Pendik Waterfront Urban Transformation Project – Istanbul – Turkey is managed by Zaha Hadid Architects. The project could be considered one of the first parametric urban projects in which generative design was included in all its design stages, starting from the concept (where the journey began to search for a methodology that simulates the existing as a reality and tells the era and its requirements) up to the final product.

The main objective of the project is to transform this part of the city from an industrial center to a commercial center and targeted for being an important focus point in the Asian part of Turkey.

4.1. Project Location

The project site is located within the Istanbul metropolitan area on the Asian continent, as in Figure 9, east of the historical and commercial center of Istanbul and about 20 km from the Bosphorus Strait.

The site project covers a total area of 550 hectares, of which 360 hectares are in the development area, and the remainder will continue within the existing greenbelt. The site is an extended piece of land, usually six hundred meters wide (approximately from east to west) and 3 kilometers long (approximately from south to north) (Zaha Hadid Architects, 2009).

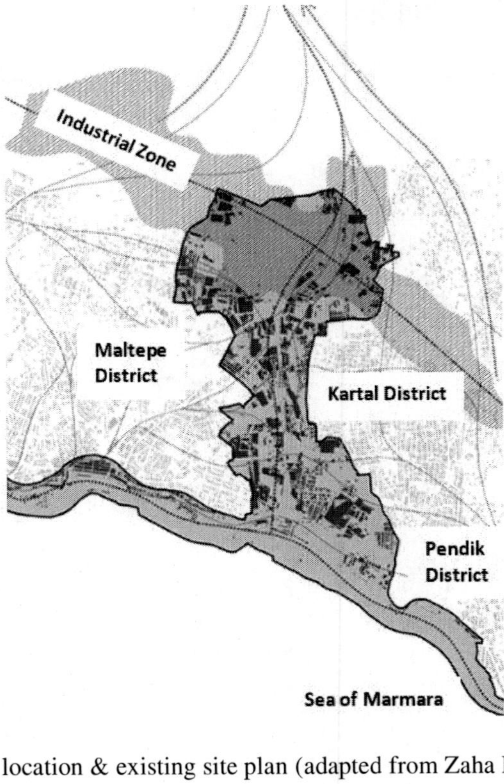

Figure 9. Project location & existing site plan (adapted from Zaha Hadid Architects, 2009).

4.2. Project Components

Commercial and service areas, hotels, tourist accommodations, housing landings, and administrative buildings. At the start of the design idea, project planning principles were considered, including Flexibility, Equality, Accessibility, Usability, Functionality, and Feasibility. These main principles served as the starting point for distributing 1,298 plots of land in the project area, with the distribution of services within walking distance and the interest in providing mass transportation lines and stopping points. (Kocabas, 2010).

4.3. Design Concept

It is based on the concept of flexibility and adaptability to the current context and on the idea of controlling the non-built rather than the built. In general, the project started with the concept of Dynamic Stitching to link the site of the master plan to the surrounding context, as in Figure 10.

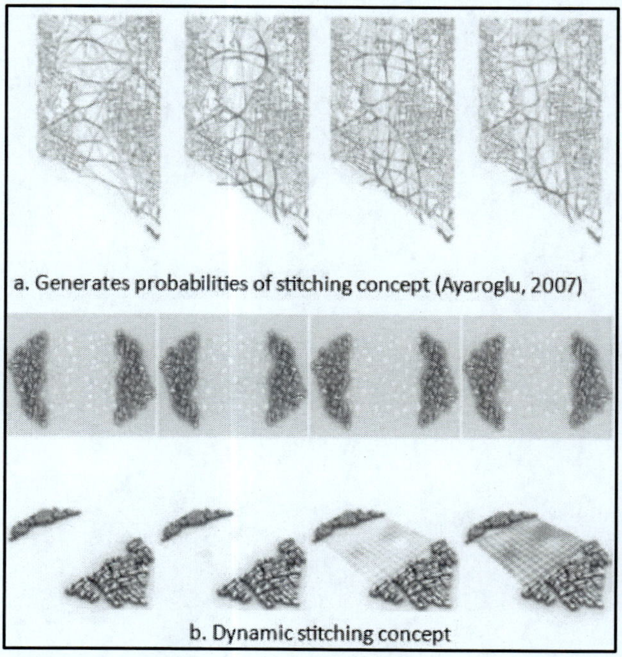

Figure 10. Steps of a concept layer generation with the existing site plan (Zaha Hadid Architects, 2009).

4.4. Stages of Project Submission

There are four levels for the project submission: information, developing, building, and optimizing level.

i. Information level: At this stage, the project collected background information such as geographical location, topographical points, land uses, and ownership of plots of land., which in turn fell into three types (regions) they are major industrial areas, public sector lands, small residential properties, as figure 11 shows. Then, Political and planning decisions were taken to transition the city from industrial to economic use., which is the concentration of economic activity in the European part and the existence of one vital center for the city.

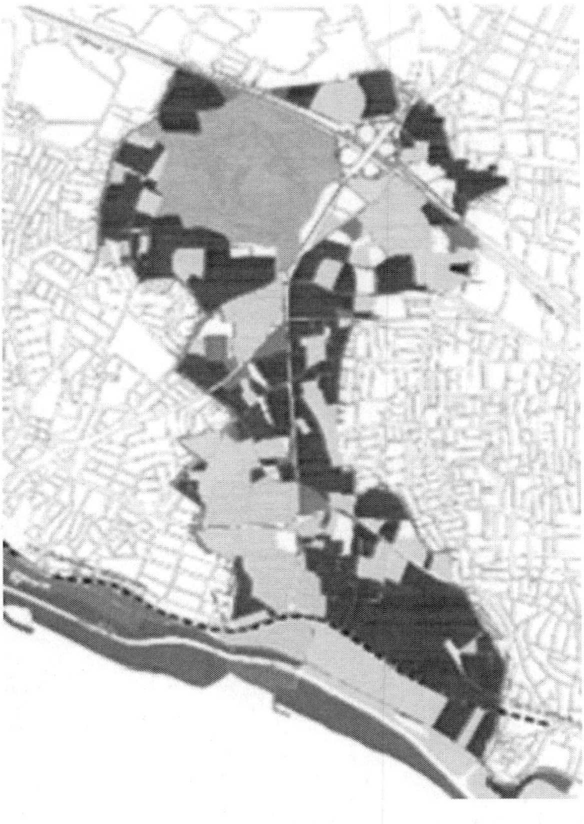

Figure 11. Information level: land use as existing (Zaha Hadid Architects, 2009).

ii. Developing Level: In addition to the main street network, they used a series of sub-road networks, which are the beginning of implementing and moving the approved design concept, which is a series of lines that move in response to the (landscape) and have different scales. These lines evolve in the different design stages, and it is a sub-network of a main network that serves as a trace of (the flow of people, the borders of the height areas, the building line, and the vision line), which together form the flexible landscape.

iii. Generate grid with 2D and 3D: Create a smooth grid that forms the basic framework for the project. In certain areas, the grid rises to form a network of towers in an open landscape, while in others, it swoops down to become a denser structure that cuts through streets, and at other times, it may fade away entirely to generate open spaces and parks. Several principles were emphasized when distributing road networks, as shown in Figure 12, which are:

   A. Connectivity: It was used at several levels and with objectives related to linking the sub-network with the existing (road networks, public spaces network, and residential areas).
   B. Continuity: to create interconnection with the existing commercial uses in parallel with the pedestrian movement. The interconnection is achieved by providing new commercial paths that are interconnected with the existing commercial areas along the main roads, thus achieving continuity of the land use.
   C. Intensify: To form high-density areas and provide open spaces, zoning, and buildings perpendicular to the city fabric.

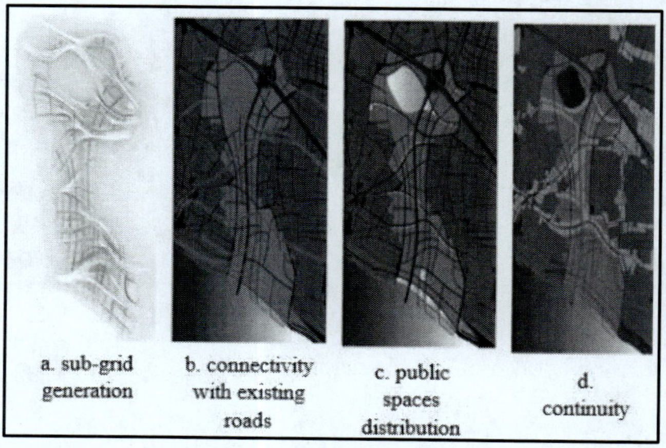

Figure 12. Developing Level (Zaha Hadid Architects, 2009).

4.5. Building Level

On the level of distributing the blocks, the starting idea was Stitching, where its strategy was used in the movement and reflected on the heights of the buildings, as it starts from a point and rises to the highest points in the central areas, achieving in those areas a higher density, as shown in Figure 13. The determination of these heights in the buildings was carried out using a 3D surface that contains a set of points (determining these points according to the land concessions) that rise with them, and the urban form moves towards them according to the relationship of mass and space.

When defining the urban neighborhoods, reference was made to the natural differentiation of the site, which divided the site into three broad ranges (the northern axis, the waterfront, and the backbone connecting them, which is characterized by the presence of a central public space and a tram) extending from west to east. Its role is not limited to distribution but is a strong urban generator to form a main pole. At this stage, parameters related to (skyline, building height, distance between buildings, vertical building line, and human scale) were also set. As for the (landscape) level, its distribution resulted from (sub-grid movement, massing distribution, pedestrian movement, and 3D surfaces).

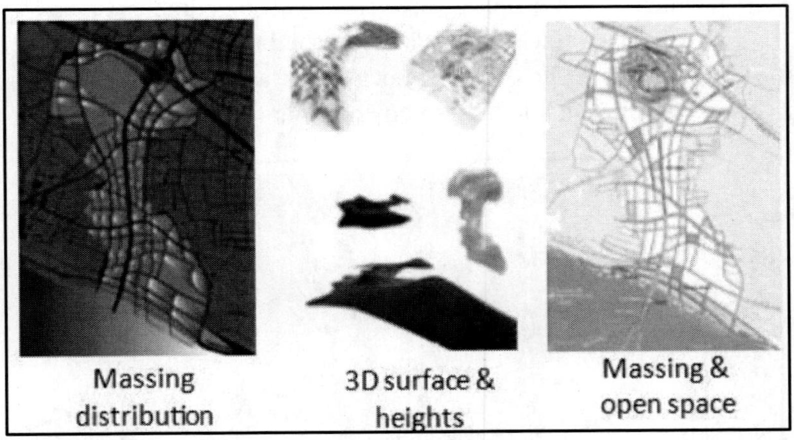

Figure 13. Building Level (Alp, Julide & Alp, İbrahim, 2014) https://www.zaha-hadid.com/masterplans/kartal-pendik-masterplan/.

4.6. Optimizing Level

At this stage, the master plan is refined and developed according to specifics related to:

Sustainability: - (Transport, water saving technologies, energy provision renewables, energy provision: high-efficiency thermal system, combined heating power generation, sustainable urban designs)

- Global eco footprinting: Reduce the use of global eco footprinting per hectare
- Transport Systems: roads, railroad, sea transportation, airway transportation.
- Resource Sycle: reduce demand, increase efficiency, use renewable energy.

In general, the region's development has moved towards adapting to the urban environment as it responds to the different requirements of each region. With linear bands around the buildings, they create open conditions that can transform from separate buildings into perimetrical blocks while retaining their character as a porous, interconnected network of open or meandering spaces throughout the city. Through diversions and gradients from one part of the site to the other, the scripted texture can create a smooth, light transition from the surrounding context represented by the neighborhoods to the new, high-density development. The flexible grid also has the potential to accommodate urban growth, as shown in Figure 14.

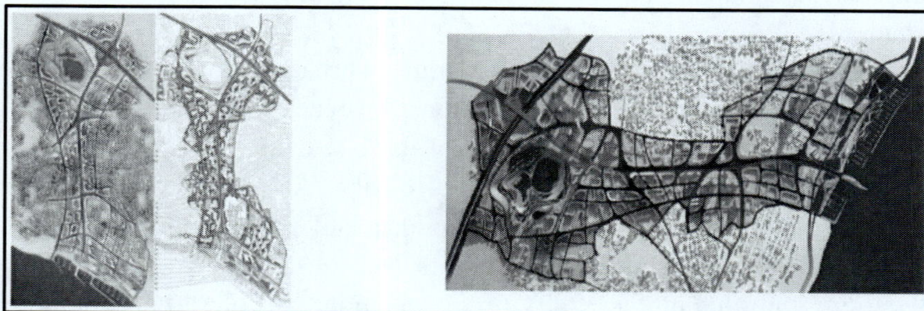

Figure 14. Final master plan (Kocabas, 2010).

4.7. Discussion

The case study represents an application of generative systems in generating urban patterns. After tracking the different design stages, it was concluded that the approach used was combining the approaches in (Paragraph 7) to reach the project goal. In addition to the possibility of different scenarios that simulate the requirements of the project and its neighbor, with the possibility of generating several different possibilities according to the inputs used, the result of the project has moved from determinism in traditional urban design to probability in contemporary urban design. In general, the master plan of the project, as explained, is a dynamic system that generates a framework that is adaptable to the urban form to different conditions. It acts as an urban text that generates different types of buildings that respond to the specific requirements of each area while leaving the possibility of continuous modification. The network was used to generate the urban form, and at other times it adheres to the characteristics of porous and interconnected with both open or winding spaces and distributing blocks throughout the city. Through different transformations and gradients (scale and shape) from one part of the site to the other, the written fabric can create a smooth, light transition from the surrounding context of the neighborhoods to the new high-density development. The flexible network also includes the possibility of accommodating urban growth. The project's development in its various stages in the environmental direction was towards being a flexible network that achieves adaptable responses to the environment. However, it is difficult to evaluate the possibility of access to this adaptation because there are still unpredictable things, such as how the existing one is adapted to environmental changes.

Knowing the role of generative systems in urban design requires studying the overlap between the stages of generative design and the process of generating an urban pattern. The role of these systems is linked to being an assisting tool in the different phases of design. As presented in the (Case Study), the designer's role is to introduce and provide the system with sufficient information and different design decisions, starting with information regarding the site and the reality of the situation, reaching to generating the final form, and submitting it to refinement according to the main principles developed by the designer and the visions of the various competent authorities. Even in the principle of the operation of generative algorithms based on parameters, their introduction requires a return to standards, the planning policy of the region, and in sustainable projects, a return to the various sustainability requirements and other determinants that move the design idea to the level of being a master plan that achieves the goals to be reached in the project.

Conclusion

The generative systems Moving urban pattern from determinism to probability in contemporary methods, in addition to moving from a specific approach to integrating the approaches that adopt (the top-down principle) and (the bottom-up principle), both need a single conceptual framework, as explained in (Design stages). These transitions only respond to the idea that contemporary urban patterns always change to adapt to contemporary and ongoing changes. There is still a major feature that cannot be ignored in generative systems that distinguish them from traditional ones. Any parameter changes lead to the adaptation of different design parts and a simulation of that change. In addition, generating systems are a tool to generate urban patterns, not to generate them by themselves, improving the quality of the selected plan. The designer remains the one who drives those. The tool is geared towards design requirements rather than vice versa.

For all disciplines in which the growth of urban cities is linked, the generative design approach helps to achieve interdependence between different disciplines, ending the idea of going through separate episodes and moving to the linked cycle between them. We are now in the phase of the digital revolution in which artificial intelligence must be invested, which simulates the understanding of architecture and deciphering its complexity. AI must be invested in to simulate the understanding of architecture and decipher its complexity, and its tools must be used to reduce cost and time. Also, moving the creative processes to a deeper stage by investing in contemporary capabilities. At some level, generative design tools will help move from Cities with urban problems to smart and sustainable cities, stimulating the growth of cities and avoiding urban mistakes. With the advancement of technology and contemporary life, there is always a need to return to the point from which urban theorists and designers ended and make it a starting point for another stage with the possibilities of that stage. Technology should be used to link contemporary cities to urban theories and avoid falling into their problems through the simulation capabilities provided by technology. Future studies should focus on the user's needs as a determinant of contemporary design while facing difficult challenges. The user has become a consumer of the environment and its resources, and the study can be used to guide them toward preserving the ecosystem with the help of generative systems.

References

Abussaud, Z. L., & Alshuwaikhat, M. H. (2023). Generative Urban Design for Smart, Sustainable Resilience Cities. Resilience Through Impact Assessment and Leadership. In *Proceedings of International Association for Impact Assessment* (IAIA2023), 8-11 May, Kuching, Malaysia.

Alexander, C., and Ishikawa, S. (1977). *A Pattern Language: Towns, Buildings, Construction*. Oxford University Press, UK.

Al-Kazzaz. (1999). *Stereotypes in Postmodern Architecture, Specificity Architect Aldo Rossi.* Master thesis, Architecture Department, University of Technology-Iraq.

Alp, Julide & Alp, İbrahim. (2014). Participation and Sustainability in Urban Projects: Istanbul Case. In *Proceedings of International Congress on Urban and Environmental Issues and Policies*, May 29-31, Akaray, Turkey.

Al-Sayed, K. (2013). Synthetic Space Syntax: A Generative and Supervised Learning Approach In Urban Design. In *Proceedings of the Ninth International Space Syntax Symposium.* Edited by Y O Kim, H T Park, and K W Seo, Seoul: Sejong University.

Ayaroğlu, M. (2007). *Urban Complexity And Connectivity: Emergence Of Generative Models In Urban Design*. Master Thesis, Middle East Technical University, Baghdad, Iraq.

Beirão, J., and Duarte, J. (2005). Urban Grammars: Towards Flexible Urban Design. In *Proceedings of the 23rd Int. eCAADe Conference*, September 21-23. Lisbon, Portugal: 491–500. https://doi.org/10.52842/conf.ecaade.2005.491.

Boumaraf, H. (2022). *A System Proposal for Generative Design Processes and Space Analysis in Architecture*. PhD. Thesis, Anadolu University Institute of Graduate Programs.

El-Zanfaly, D. (2009). Design by Algorithms: A Generative Design System for Modular Housing Arrangement. In *Proceedings of the 13th Congress of the Iberoamerican Society of Digital Graphics*, (SIGraDi 2009) Sao Paulo, Brazil, November 16-18, 2009.

Fernandes, S. R. M. (2013). *Generative Design: A New Stage in the Design Process*. [Master thesis]. Public University in Lisbon, Portugal.

Frazer, J. (2002). Creative Design and the Generative Evolutionary Paradigm. Chapter 9 in the book *Creative Evolutionary Systems*, Elsevier: 253-274. https://doi.org/10.1016/B978-155860673-9/50047-1.

Gashenko, A. (2019). Pattern method in urban studies and practices. *Architecture and Engineering*, 4(2), 33–39. https://doi.org/10.23968/2500-0055-2019-4-2-33-39.

Ghusoon Najm, & Al-Yousif, I. J. K. (2021). *The Logic of Ideas in Urban Design. Source*? Assignment report, Architecture Engineering Department, University of Technology, Baghdad, Iraq. https://doi.org/10.13140/RG.2.2.17562.57283.

Gradišar, L., Klinc, R., Turk, Ž., & Dolenc, M. (2022). Generative Design Methodology And Framework Exploiting Designer-Algorithm Synergies. *Buildings*, 12(12), 2194. https://doi.org/10.3390/buildings12122194.

Han, Z., Yan, W., & Liu, G. (2021). A Performance-Based Urban Block Generative Design Using Deep Reinforcement Learning and Computer Vision. In P. F. Yuan, J. Yao, C. Yan, X. Wang, & N. Leach (Eds.), *Proceedings of the 2020 Digital FUTURES*. Springer Singapore: 134–143. https://doi.org/10.1007/978-981-33-4400-6_13S.

İnceoğlu, M., Anadolu Üniversitesi, F. B. E., & Boumaraf, H. (2022). *A System Proposal for Generative Design Processes and Space Analysis in Architecture*. PhD. Thesis, Anadolu Üniversitesi - Fen Bilimleri Enstitüsü. https://earsiv.anadolu.edu.tr/xmlui/handle/11421/27101.

Jacobs, J. (1961). *The Death And Life of Great American Cities*. Random House, New York, USA.

Jović, B. S., & Mitić, A. D. (2020). Exploration of Nature-Based Biomimetic Approach in Landscape Architectural Design: Parametric Study of Candelabra Model Design. *Visual Computing for Industry, Biomedicine and Art,* 3(1): 25. https://doi.org/10.1186/s42492-020-00060-y.

Kocabas, Arzu. (2010). Kartal Urban Regeneration Project: Challenges, Opportunities, and Prospects for The Future. *WIT Transactions on Ecology and the Environment*, 129: 571-582. http://dx.doi.org/10.2495/SC100481.

Koenig, R. (2018). *Generating Urban Structures: A Method for Urban Planning Supported by Multi-Agent Systems And Cellular Automata*. OSF Preprints s4crf, Center for Open Science. https://doi.org/10.31219/osf.io/s4crf.

Koenig, R., and Bauriedel, C. (2004). Computer-Generated Urban Structures. In *Proceedings of 7th Generative Art Conference GA2004*. 14-16 December 2004. University of East London, London, UK. https://doi.org/10.13140/RG.2.1.1259.1847.

Kumalasari, D., Koeva, M. N., Vahdatikhaki, F. Petrova-Antonova, D., and Kuffer, M. (2022). Generative Design for Walkable Cities: A Case Study of Sofia. *The International Archives of the Photogrammetry, Remote Sensing and Spatial Information Sciences*, XLVIII-4/W5-2022: 75–82. https://doi.org/10.5194/isprs-archives-XLVIII-4-W5-2022-75-2022.

Mandić, M., and Tepavčević, B. (2015). Analysis of Shape Grammar Application As A Tool For Urban Design. *Environment and Planning B: Planning and Design*, 42(4): 675–687. https://doi.org/10.1068/b130084p.

Marshall, S. (2005). *Streets and patterns* (1st ed). Routledge, London, UK. https://doi.org/10.4324/9780203589397.

Orabi, R. (2016). *The Role of Computation in the Architectural Design Process*. Master Thesis, Damascus University, Syria.

Qeisi, S. A., and Al-Alwan, H. (2021). Generative Urban Design Concepts And Methods: A Research Review. *IOP Conference Series: Materials Science and Engineering*, 1090(1): 012085. https://doi.org/10.1088/1757-899X/1090/1/012085Mandić.

Salman, M., and Rasheed, K. G. (2018). The Influence of the Urban Pattern Relationship With The Architectural Model On The Integration Of The Formal Structure Of The Urban Centers. *Journal of Planner and Development Center of Urban & Regional Planning*, 23(1): 22-49.

Shehu, M., and Yunitsyna, A. (2017). Generative Computational Tools for the Design of Urban Morphology. *Fioravanti, A, Cursi, S, Elahmar, S, Gargaro, S, Loffreda, G, Novembri, G, Trento, A (Eds.), Sharing Computational Knowledge, Proceedings of the 35th eCAADe Conference* 1, Sapienza University of Rome, Rome, Italy, 20-22 September: 477-482. http://papers.cumincad.org/cgi-bin/works/paper/ecaade2017_118.

Stavric, M., and Marina, O. (2012). Application of Generative Algorithms in Architectural Design. *Advances in Mathematical and Computational Methods*, 37: 175-180.

Terzidis, K. (2006). *Algorithmic architecture* (1st ed). ISBN: 9780750667258. Routledge, Taylors and Francis group. Amsterdam, Netherlands.

Westerveld, C. E. (2021). *Generative Design: Recommended Actions to Smooth the Way For Production Of Generative Designs With Additive Manufacturing.* Master Thesis, University of Twente, Netherlands. http://essay.utwente.nl/85976/.

Zaha Hadid Architects. (2009). *Kartal masterplan, istanbul* [Final Report]. http://www.kartalkentder.org/upload/Node/38715/files/Kartal-Masterplan_.pdf.

Zuckar, P. (1959). *Town and square from the agora to the village green.* Columbia University Press, New York. https://archive.org/details/townsquarefromag0000zuck/page/n1/mode/2up.

Chapter 4

Phase Noise Compensation in Optical Fiber Based on Filter Bank Multi-Carrier in 5G Applications

Ahmed H. Abbas[1,*]
Thamer M. Jamel
and Jabbar K. Mohammed
[1]University of Technology- Iraq, Baghdad, Iraq.

Abstract

This chapter discusses the advancement in the 5G communication technology by filter bank multicarrier technique to solve the problem of phase noise in optical fiber systems. To make 5G a reality, a waveform that provides better spectral confinement per subcarrier than orthogonal frequency division multiplexing is needed. Filter bank multi-carrier fits the bill. This research aimed to examine existing methods of phase noise compensation for optical fiber systems that use filter bank multicarrier. Four distinct phase tracking methods have been considered in the context of a single input single output filter bank multicarrier–offset quadrature amplitude modulation system over an optical channel: maximum likelihood, maximum a posteriori, adaptive maximum likelihood, and modified blind phase search. Each algorithm's strengths and weaknesses are analyzed so that their capabilities and performance can be better understood. An outlook on forthcoming studies is also provided, aiming to improve the efficacy of phase noise compensation techniques in filter bank multicarrier-based optical fiber systems.

* Corresponding Author's Email: coe.21.01@grad.uotechnology.edu.iq.

In: Advancements in Sustainability Systems
Editors: Hussain H. Al-Kayiem, Hasanain A. Abdul Wahhab and Raed A, Jessam
ISBN: 979-8-89113-571-0

Keywords: maximum likelihood, maximum a posteriori, adaptive maximum likelihood, modified blind phase search, filter bank multicarrier

Abbreviations

AFB	Analysis filter bank.
CD	Chromatic dispersion.
CPE	Common phase error.
DWDM	Dense wavelength division multiplexing.
DPSK	Discrete phase shift keying.
FBMC	Filter bank multicarrier.
ICI	Intercarrier interference.
MBPS	Modified blind phase search.
NMSE	Normalized mean square error.
OQAM	Offset quadrature amplitude modulation.
OFDM	Orthogonal frequency division multiplexing.
PN	Phase noise.
SFB	Synthesis filter bank.
SNR	Signal to noise ratio.

Nomenclature

$r[n]$	received signal.
$2M$	subcarrier number.
Δv	Laser linewidth.
S(f)	transmitted signal.
$d_{m,l}$	Real valued multicarrier symbol (SFB input) at subcarrier M and multicarrier symbol I.
$\tilde{Z}_{m,l}$	Demodulated symbol (SFB input) at subcarrier M and multicarrier symbol I.
$\tilde{z}^{R}_{m,l}$	Real demodulated symbols (AFB outputs) at subcarrier m and multicarrier symbol l.
σ_{ϵ}^{2}	Zero mean-variance
ϵ_l	Estimation symbol
φ_l	Current phase
$\hat{\varphi}_{l-1}$	Previous phase
$\omega[n]$	Additive noise.

1. Introduction

One of the most modified orthogonal frequency division multiplexing (OFDM) is filter bank multicarrier (FBMC), which is the development of OFDM that makes specific improvements. Still, it requires cost and more complex signal processing (Alper, 2007), which can be considered an issue of FBMC. Because of their efficiency and resilience, FBMC systems are being evaluated as a potential replacement for future communication standards, notably 5G (Gerzaguet et al., 2017). FBMC uses a sine prototype filter. The primary lobe width of the sine prototype filter in the frequency domain is three times the sub-carrier spacing, but in the OFDM system, it is only two times. The improved spectral decay and a larger main lobe of FBMC are better than OFDM regarding carrier frequency offset sensitivity (Davide et al., 2021), as shown in Figure 1 and Figure 2.

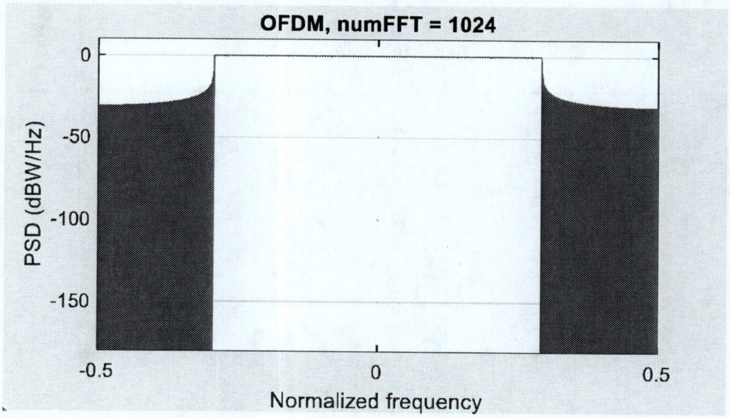

Figure 1. Power spectral density of orthogonal frequency division multiplexing.

Due to the elimination of cyclic prefixes and narrower guard bands between frames (Kansal et al., 2000), filter bank multicarrier (FBMC) is another approach to surpassing the constraints of OFDM systems. Spectrum leakage is minimized in FBMC compared to cyclic prefix OFDM systems (Schellmann et al., 2014). FBMC reduces interference between subcarriers in seconds, making the system more reliable and long-distance contact more dependable. Pinpoint many users with varying center frequencies or spectral gaps (Gopal et al., 2015). Figures 3 and 4 show the difference between OFDM and FBMC. Detection and cancellation of multiple access interference (MAI) necessitates the implementation of receiver synchronization. The capacity of subcarriers to accurately determine their frequencies enables the reduction of multiple access interference (MAI) (Bidyalaxmi et al., 2014).

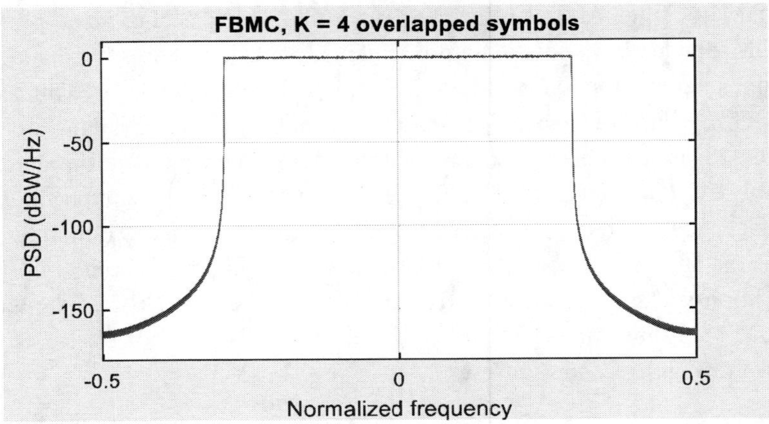

Figure 2. Power spectral density of filter bank multicarrier.

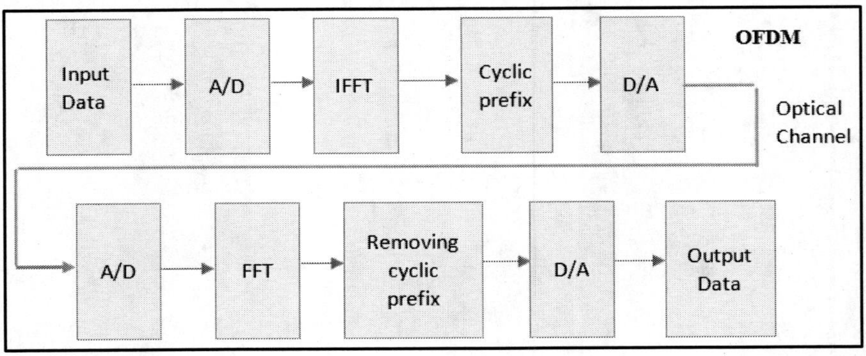

Figure 3. Orthogonal frequency division multiplexing physical layer block diagram.

However, at such high frequencies, phase noise (PN) can significantly affect the performance of the communication system (Khanzadi et al., 2014). FBMC is more sensitive to the PN effect. The generated MMSE-based estimated channel will be used with pilot symbols during the data transmission phase to account for PN and identify the broadcasted data. The PN vector must be expressed in terms of known basis vectors before it can be determined and corrected. PN is a particular issue for multi-carrier systems due to their physical constraints. Phase Noise's impact on digital data modulation has been studied for several decades (Xu et al., 2016).

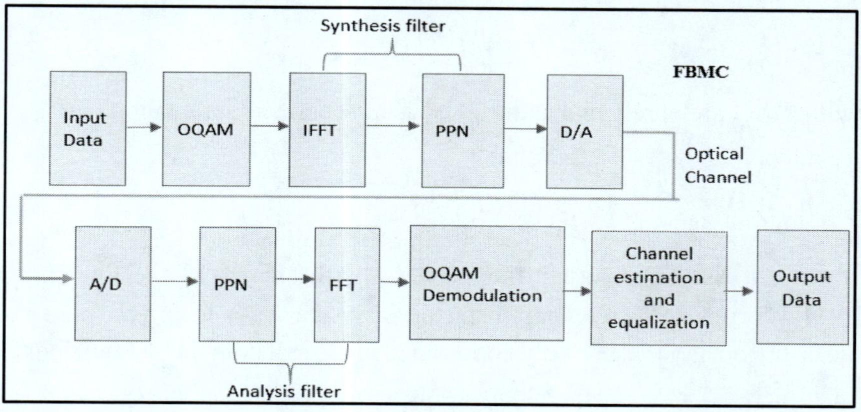

Figure 4. Filter bank multicarrier physical layer block diagram.

The impact of Phase Noise may be seen as a scattering noise in the form of a composite exponential function in the time domain (Zhang et al., 2021). However, the Phase Noise effect may be decomposed into the common phase error (CPE) (Khattab et al., 2017). And the intermodulation cancellation (ICI).

The effects of phase noise (PN) on communication system performance are the subject of this essay, with Filter Bank Multicarrier (FBMC) systems serving as a case study. We discuss how vulnerable FBMC is to PN effects and why we need reliable ways of correction. This article suggests and describes a new scheme for compensating phase noise (PN) using trackers such as ML, MAP, AML, and M-BPS in FBMC/OQAM optical systems.
The FBMC/OQAM optical system model is presented in Section 2. Current approaches to PN compensation are discussed in Section 3. Previous research on phase noise is presented in Section 4. Finally, a new strategy for compensating for PN is proposed in Section 5. The chapter conclusion is wrapped up in section 6.

2. FBMC-OQAM Optical System Model

The FBMC-OQAM system, shown in Figure 5, focuses on the PN impact, considers only one polarization mode, and assumes that the chromatic dispersion (CD) is perfectly compensated at the receiver. NT and NR denote the number of transmitted and received signals; the transmit signal for an FBMC-OQAM system is given by (Trung et al., 2017), sing Equation (1).

$$s(n) = \sum_{m=0}^{2M-1} \sum_{l=0}^{2N_s-1} d_{m,l} p_{m,l}(n) \tag{1}$$

where, $d_{m,l}$ are purely real and $p_{m,l}(n)$ is now defined in Equation (2):

$$p_{m,l}(n) = j^{m+l} p(n - lM) e^{j\frac{\pi m}{M(n-(L_p-1)/2)}} \tag{2}$$

where *p(n)* is the prototype filter and L_p Is the length of the filter, $L_p = 2KM$. Where K is the overlapping factor. Since the symbols $d_{m,l}$ are pure real, the orthogonality must be only constrained in the real domain by Equation (3).

$$\Re\left(\sum p_{m,l}(n) p^*{}_{m0,l0}(n)\right) = \delta_{m-m0,l-l0} \tag{3}$$

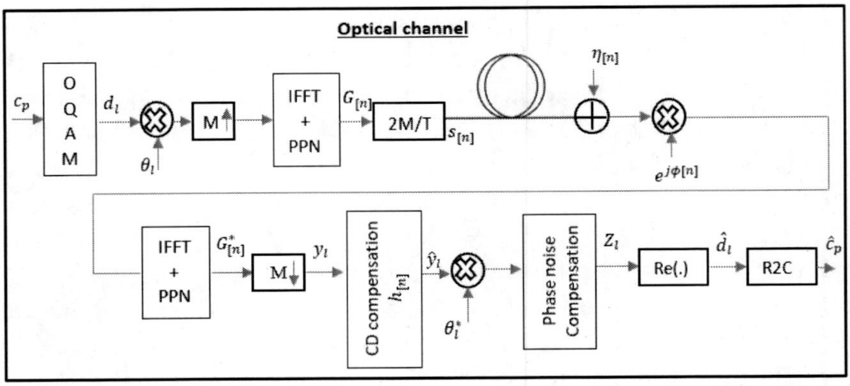

Figure 5. FBMC-OQAM transmission chain.

The received signal in Equation (4) is only impacted by PN, and additive noise is given by:

$$r[n] = s[n]e^{j\varphi[n]} + w[n] \tag{4}$$

The PN is modeled through equation 5 (François et al., 2014):

$$\varphi[n+1] = \varphi[n] + \nu[n] \tag{5}$$

ν[n] is a Gaussian variable selected randomly, provided that it is real and individualistic and has a variance of $\sigma_\nu^2 = 2\pi\Delta\nu\frac{T}{2M}$, Δν is the combined laser linewidth, the baseband received signal $r[n]$, impacted by additive noise $\omega[n]$. As S(f) the Fourier transforms of the transmitted signal, PN may be expressed by equation 6.

$$r[n] = \left(\int_{\frac{-1}{2T_S}}^{\frac{1}{2T_S}} S(f)\, H(f)\, e^{j2\pi f T_S n}. df \right) e^{j\emptyset[n]} + \omega[n]. \tag{6}$$

At the subcarrier level, the channel has to be frequency flat, and the PN has to be slowly varied in FBMC-OQAM systems. In such a case, if the discrete receive prototype filter g[n] pulse specified in frequency and time, get z_{m_0,l_o}as the neglecting additive noise by Equation 7:

$$z_{m_0,l_o} = \sum_{n=0}^{2M\,k-1} r[n] g^*_{m_0,l_o}[n] \approx$$
$$e^{j\emptyset l_0} H_{m_0} \sum_{l=-\infty}^{+\infty} \sum_{m=1}^{2M-1} d_{m,l} \sum_{n=0}^{2M\,k-1} p_{m,l}[n] g^*_{m_0,l_o}[n]. \tag{7}$$

H_m =$H_m(f_m)$; f_m is the frequency of subcarrier m and can obtain the transmitted symbols back by taking the real part d_{m_0,l_o} in Equation. (8):

$$\hat{d}_{m_0,l_o} = \Re \left\{ \frac{z_{m_0,l_o}}{e^{j\emptyset l_0} H_{m_0}} \right\} \tag{8}$$

Phase noise (PN) is a special case of time selectivity that causes a phase rotation of the received samples (Tao et al., 2009) and is modeled by Eq. (1). The phase noise issue is essential to the system's effectiveness because phase noise results in inter-carrier interference (ICI), which rotates the constellation map and degrades the receiving signal quality. Phase noise originates from the lasers' phase mismatch between the transmitter and the receiver (Trung et al., 2017).3.

Phase Noise Compensation Methods

Phase noise compensation methods on a single input single output (SISO) are outlined in this section. FBMC-OQAM system uses an optical channel with four different tracking algorithms, namely, ML, MAP, AML, M-BPS) (François et al., 2017). Also, the main reasons for phase noise in optical fiber are Chromatic dispersion, polarization mode dispersion, Rayleigh scattering, Brillouin scattering, and Raman scattering caused by different wavelengths of light traveling at different speeds through the fiber. Hence, there are four methods to estimate the noise in optical fiber (OP):

3.1. Maximum Likelihood (ML)

Consider the estimation of the previous phase at multicarrier symbol $l-1$ by$\hat{\varphi}_{l-1}$. The estimation of the current phase. φ_l denote by$\hat{\varphi}_l$ Given that the PN is expected to fluctuate slowly throughout a symbol, we may estimate the phase error by substituting a current estimate for the unknown one; let us donate. $\epsilon_l = \varphi_l - \hat{\varphi}_{l-1}$ and its estimation, $\hat{\epsilon}_l = \hat{\varphi}_l - \hat{\varphi}_{l-1}$, instead of re-estimating φ_l. The ML estimator of ϵ_l is the one that maximizes the likelihood of the demodulated symbols given (Xinwei et al., 2020). based on currently available decisions, using Equation (9).

$$\hat{\epsilon}_l^{ML} = \arg\max_{\epsilon_l} f(\tilde{z}_{0,l}, \dots, \tilde{z}_{2M-1,l}|\epsilon_l) \tag{9}$$

3.2. Maximum A Psteriori (MAP)

The ML estimator does not make use of the a priori distribution of ϵ_l. In actual use, both the sending and receiving lasers are known. Therefore, the linewidth can be estimated, and we can refine the estimate with this information. The MAP estimator of ϵ_l is given by Equation (10).

$$\hat{\epsilon}_l^{MAP} = \arg\max_{\epsilon_l} f(z|\epsilon_l) f(\epsilon_l) \tag{10}$$

Assuming that the previous phase estimate is close to the actual phase, we can approximate it. $\epsilon_l = \varphi_l - \hat{\varphi}_{l-1}$ by $\epsilon_l = \varphi_l - \hat{\varphi}_{l-1}$ Under this assumption, $\epsilon_l$is normally distributed with zero mean and variance in Equation (11)

$$\sigma_\epsilon^2 = M\sigma_v^2 = \pi\Delta v T, i.e., \tag{11}$$

$$f(\epsilon_l) = \frac{1}{\sqrt{2\pi\sigma_\epsilon^2}} \exp\left(-\frac{\epsilon_l^2}{2\sigma_\epsilon^2}\right), \tag{12}$$

$$\log f(\epsilon_l) = D - \frac{\epsilon_l^2}{2\sigma_\epsilon^2}, \tag{13}$$

where D in Equation (13) is a constant independent of ϵ_l.

3.3. Adaptive Maximum Likelihood (AML)

Adaptive maximum likelihood (AML) is a statistical technique used to estimate the parameters of a statistical model (Rottenberg, 2018). AML is a variation of the maximum likelihood estimation (MLE) method, which seeks to find the parameter values that maximize the likelihood function of the data given the model, as shown in Equation (14). Assuming that we are in a low-noise scenario (σ_u^2 >> N0),s and AML PN estimator is obtained as:

$$\hat{\phi}_l + 1 = \hat{\phi}_l + \mu \sum_{m=0}^{2M-1}(\tilde{z}_{m,l}^I cos\hat{\epsilon}_l - \tilde{z}_{m,l}^R sin\hat{\epsilon}_l)(d_{m,l} - \tilde{z}_{m,l}^R cos\hat{\epsilon}_l - \tilde{z}_{m,l}^I sin\hat{\epsilon}_l) \quad (14)$$

where $\hat{\epsilon}_l$ was defined in Equation (15) as:

$$\hat{\epsilon}_l = \hat{\phi}_l - \hat{\phi}_{l_{-1}} \cdot \quad (15)$$

μ: is a step size parameter controlling the convergence rate, which needs fine-tuning to function at its highest potential. Also, the complexity of AML is lower than that of M-BPS.

3.4. Modified-Blind Phase Search (M-BPS)

M-BPS extends the traditional blind phase search (BPS) technique (Wu et al., 2019), which uses a phase rotator to iteratively rotate the received signal to align the carrier phase with a reference phase. However, BPS is computationally intensive and can require many iterations to converge to the correct carrier phase. When applied to FBMC/OQAM optical fiber systems, the blind phase search technique can improve performance and reliability, making them more suited for high-speed data transmission (Nguyen et al., 2016).

4. Related Studies

When a signal's phase varies or fluctuates at random, this is called phase noise. Phase noise is a source of degradation in the performance of communication systems. There are two primary components:

Common Phase Error (CPE) and Intercarrier Interference (ICI). Phase noise includes common phase error (CPE), which occurs when all the system's subcarriers experience the same systematic phase offset. Inaccuracies in the frequency reference or system components' flaws can also contribute to this phenomenon. The entire signal can be shifted or distorted when CPE is present, causing data detection mistakes and lowering the system's performance.

Phase noise also includes intercarrier interference (ICI), distinct from intercarrier phase rotation. ICI occurs not within a single subcarrier but due to interference from nearby subcarriers. This interference can cause signal degradation and poor system performance by introducing distortions and overlaps between adjacent subcarriers. Lin, 2018 reported a description of the effects of phase noise on system performance, which includes CPE and ICI. The results of these factors on the system and their precise effects on signal quality and data transfer are elaborated upon in the source cited.

Figures 6 and 7 visually represent the distinction between intercarrier interference (ICI) and phase rotation before and after compensation. This diagram depicts the interference patterns brought on by intercarrier interference (ICI), which occurs when neighboring subcarriers interact and impact one another. It emphasizes that, unlike intercarrier phase rotation, which refers to fluctuations in phase within a single subcarrier, ICI results from interactions between subcarriers, leading to signal distortion, overlap, and mistakes. The effects of phase noise on a communication system can be better understood by consulting the cited source (Yubin, 2018).

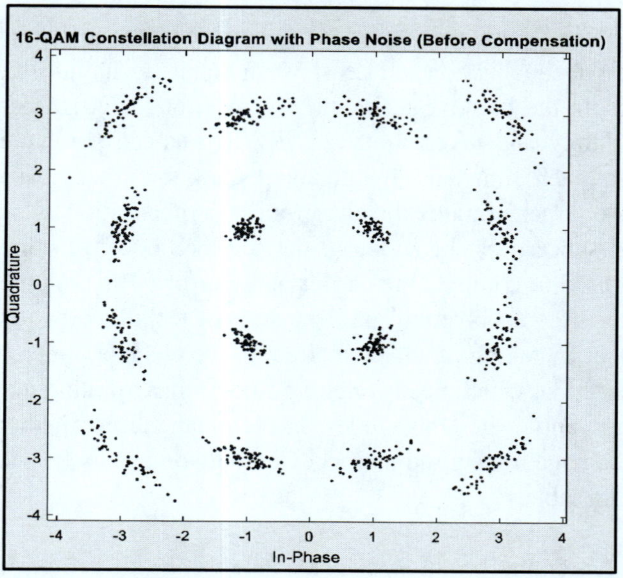

Figure 6. Phase rotation before compensation.

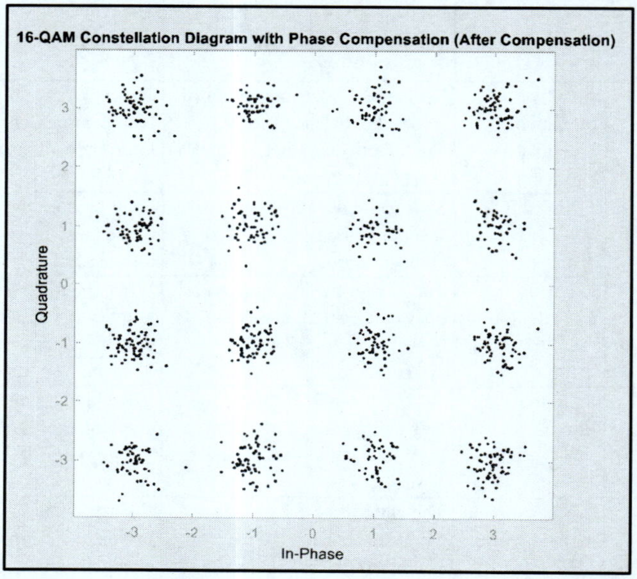

Figure 7. Phase rotation after compensation.

ML and MAP by François et al., 2017 are less influenced by additive noise since they are implemented by summing the subcarrier outputs and taking the maximum phase by argument function. When the phase noise is compensated using a one-tap equalization and Least Mean Square technique, the computing cost rises as in the iterative technique (Nguyen et al., 2016) because of the need to multiply and invert matrices. FBMC's adaptive features are made possible for each time and frequency domain by fewer bases and pilot symbols. Good performance in phase noise compensation is possible even with a few subcarriers. In addition, the log-likelihood function of the PN estimate is used to compute the carrier phase error (CPE) (You et al., 2019). This chapter uses an inverted gradient descent technique to reduce the PN estimate error by taking the inverse effect of the channel, unlike the iterative PN compensation, which needs ongoing fine-tuning by using many iterative. The system is anticipated to have less latency than current methods since it is considered a huge transmission block. Comparison results for various papers are shown in Table 1.

Table 1. Comparison of results and conclusions for various literature

Author	Technique	Advantage	Limitation
Demir, 2007	DWDM	Dispersion enhances DPSK's efficacy in the nonlinear phase noise.	The reduction in background noise is insufficient to account for.
Trung et al.,, 2017	Adaptive Maximum Likelihood Estimator	Subcarrier counts are carefully selected to balance simplicity and effectiveness in CD and PN compensation.	The high range of values for both the fiber length and the laser linewidth is still unclear.
Lin, 2018	CPE and ICI	SNR can reduce SER with or without phase noise compensation.	Fail to reduce the impact of phase noise in a high signal-to-noise ratio.
Gu et al., 2019	ICI Cancellation	The ICI caused by phase noise does not necessitate total elimination in real-world systems.	When the MCS is high, the leftover ICI will harm the system's operation.
Khaled et al., 2020	BER and Power	1-tap equalizer, N-tap equalizer getting a better bit error rate and low mean square error	ISI, which is generated by chromatic dispersion
Minkeun et al., 2022	Linear estimate method with a MMSE	The NMSEs of the various estimators used in the combined PN and channel prediction were calculated.	Estimation of the performance-limiting ICI elements in mm. Wave systems has not been a priority.
Proposed method 2023	PAFB with MBPS	Mitigate the interference between symbols	More computationally

5. Proposed Method for Phase Noise Tracking Technique

Several approaches to PN tracking in Blind feedforward PN compensation attract our attention since it does not need the employment of pilots, which degrades spectral efficiency. The modified-blind phase search (M-BPS) technique has been found to perform better than the others by Trung and Christophe, 2017). This is particularly true when the quantity of subcarriers is considerable despite its significant computing effort. The CD compensation block is the most challenging digital signal processing (DSP) step. Therefore, even while the M-BPS's complexity increases with the modulation's order, it has little impact on the complexity of the whole system (Takeo et al., 2019). As a result, we conducted this study with a new proposal method called the M-BPS technique, as shown in Figure 8. To successfully recover the sent signals, PN compensation is required. We will enhance the BPS algorithm to efficiently correct the PN in FBMC/OQAM systems by observing the constellations with and without phase noise. Remember that phase noise tracking also accounts for the little lingering carrier frequency offset (CFO).

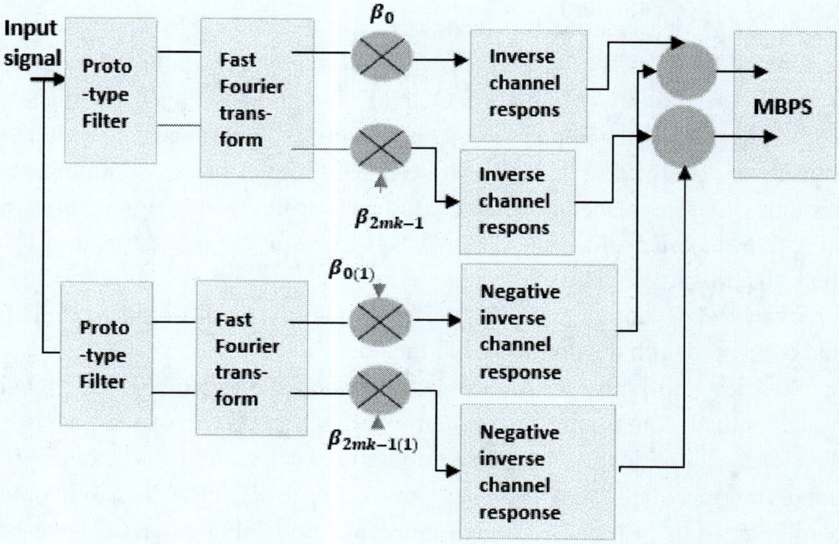

Figure 8. proposed method Parallel Analysis filter bank with M-BPS Estimator.

M-BPS is recommended for repairing the PN in the frequency domain by (Govind et al., 2017). For each symbol time, the PN is estimated using data from the analysis filter bank (AFB) output for all subcarriers. Compared to per-subcarrier PN compensation methods, the proposed approach is drastically different. M-BPS simplifies the calculation of distance in the complex plane compared to other M-BPS algorithms by just considering the distance in the real plane. Because of this (Wu et al., 2020). Fewer multiplications are required, allowing the computer to focus on other tasks. The Parallel Analysis Filter Bank (PAFB) is a flexible signal processing methodology that partitions an input signal into numerous sub-bands particular to different frequencies. This allows for simultaneous and multi-scale examination of the content inside the signal. The signal is then divided into parallel channels, and each channel has its own frequency. Also, choose the low pass filter to decrease the noise signal. Then, the MBPS is used, multiplying the received signal by the known carrier frequency reference signal, and getting the angular phase difference between the filtered and original filter.

Conclusion

Better spectral efficiency and resistance against multipath fading make FBMC-OQAM an attractive technique for optical cable communication systems. Phase noise, a problem with all communication systems, may degrade the quality of FBMC-OQAM signals and lead to mistakes. Researchers have developed various compensation approaches to reduce the impact of phase noise in FBMC-OQAM-based optical communication systems. Different methods were studied before. The best method was Adaptive maximum likelihood (AML). It had a better PN estimation result than ML, MAP, and M-BPS. A new method is proposed. It can be considered a future work to mitigate the (PN) effect where the results are accomplished and compared with the conventional methods.

Studies into phase noise compensation techniques for FBMC-OQAM-based optical communication systems are under continuous research and development by many specialized research centers and companies. It is important to compensate for phase noise in FBMC-OQAM-based optical communication systems. Although numerous promising solutions have been developed, further research is required to establish the most effective and feasible way.

References

Alper, D. (2007). Nonlinear Phase Noise in Optical-Fiber-Communication Systems. *Journal of Lightwave Technology*, 25(8): 2002-2032. https://doi.org/10.1109/JLT.2007.900888.

Bidyalaxmi, D., Nongmaithem, L., and Sonika., S. (2014). Comparative Analysis of FBMC and OFDM Multicarrier Techniques for Wireless Communication Networks. *International Journal of Computer Applications*, 100(19): 559-563. https://doi.org/10.1109/ICOEI.2018.8553689.

Daiming, Q., Fang, W., Yan, W., Tao J., and Behrouz, F. (2017). Improving Spectral Efficiency of FBMC-OQAM Through Virtual Symbols. *IEEE Transactions on Wireless Communications*, 16(7): 4204-4215. https://doi.org/10.1109/TWC.2017.2693277.

Davide, M., Mario T., and Maurice B., (2021). Comparing the Performance of OFDM and FBMC Multicarrier Systems in Doubly Dispersive Wireless Channels. *Signal Processing,* 179: 107818 https://doi.org/10.1016/j.sigpro.2020.107818.

Faulkner, M. (2000). The Effect of Filtering on The Performance of OFDM Systems. *IEEE Transactions on Vehicular Technology*, 49(5): 1877-1884. https://doi.org/10.1109/25.892590..

François, R., Trung, N., Simon, G., Jérome, L., and François, H. (2017). ML and MAP Phase Noise Estimators for Optical Fiber FBMC-OQAM Systems. *IEEE International Conference on Communications (ICC)*, Paris, France. https://doi.org/10.1109/ICC.2017.7996768.

Gerzaguet, R., Bartzoudis, N., Baltar, L. G. et al., (2017). The 5G Candidate Waveform Race: A Comparison of Complexity and Performance. *EURASIP Journal on Wireless Communications and Networking*, 13. https://doi.org/10.1186/s13638-016-0792-0.

Gopal, S. G., Metkar, S. S., and Khanchandani, K. B. (2015). Performance Enhancement of Multirate Digital Filter Structures Using Advanced DSP Optimization Techniques. In *Proceedings of the International Conference on Information Processing*, Pune, India: 307-311. https://doi.org/10.1109/INFOP.2015.7489398.

Govind, V., Mustafa, Q., Maurice, O., John, C., and Rongqing, H. (2017). Phase Noise Characterization of a QD-Based Diode Laser Frequency Comb. *Optics Express*, 25(14): 15,890-15,904. https://doi.org/10.1364/OE.25.015890.

Gu, S., Long, H., and Li, Q. (2019). Phase Noise Estimation and Compensation Algorithms for 5G Systems. In: Liu, X., Cheng, D., Jinfeng, L. (eds) Communications and Networking. China Com. *Lecture Notes of the Institute for Computer Sciences, Social Informatics and Telecommunications Engineering*, 262. Springer, Cham. https://doi.org/10.1007/978-3-030-06161-6_54.

Kansal, P., and Shankhwar, A. (2017). FBMC vs OFDM Waveform Contenders for 5G Wireless Communication System. *Wireless Engineering and Technology*, 8: (59-70) https://doi.org/10.4236/wet.2017.84005.

Khaled, A., Heng-Siong, L., and Tawfig, E. (2020). Compensation of Chromatic Dispersion and Nonlinear Phase Noise Using Iterative Soft Decision Feedback Equalizer for Coherent Optical FBMC/OQAM Systems. *Journal of Lightwave Technology*, 38(15): 3839-3849. https://doi.org/10.1109/JLT.2020.2981481.

Khanzadi, M. R., Kuylenstierna, D., Panahi, A., Eriksson, T., and Zirath, H. (2014). Calculation of the Performance of Communication Systems from Measured Oscillator Phase Noise. *IEEE Transactions on Circuits and Systems I*, 61(5): 1553-1565. https://doi.org/10.1109/TCSI.2013.2285698.

Yubin L. (2018). The Effect and Compensation of Phase Noise on Orthogonal Frequency Division Multiplexing (OFDM) System. *Circuits and Systems*, 9: 107-115. https://doi.org/10.4236/cs.2018.97011.

Minkeun, C., Liang, L., and Ove, E. (2022). Phase-Noise Compensation for OFDM Systems Exploiting Coherence Bandwidth: Modeling, Algorithms, and Analysis. *IEEE Transactions on Wireless Communications*, 21(5): 3040-3056. https://doi.org/10.1109/TWC.2021.3117782.

Nguyen, T., Gorza, S., Louveaux, J., and Horlin, F. (2016). Low-Complexity Blind Phase Search for Filter Bank Multicarrier Offset-QAM Optical Fiber Systems. *Advanced Photonics 2*, OSA Technical Digest, Optica Publishing Group, paper SpW2G.2. https://doi.org/10.1364/SPPCOM.2016.SpW2G.2.

Philip, N.J., Ting, W., Dayou, Q., Lei, X., Yoshiaki, A., Tsutomu T., Christoforos, K., and Konstantinos, K. (2012). Demonstration of High-Speed MIMO OFDM Flexible Bandwidth Data Center Network. *European Conference and Exhibition on Optical Communication*, Optica Publishing Group. https://doi.org/10.1364/ECEOC.2012.Th.2.B.1.

Rostom, Z., and Didier, R. (2016). Theoretical Analysis of the Power Spectral Density for FFT-FBMC Signals. *IEEE Communications Letters*, 20(9): 1748-1751. https://doi.org/10.1109/LCOMM.2016.2588497.

Rottenberg, F. (2018). *FBMC-OQAM Transceivers for Wireless and Optical Fiber Communications,* UCL-UniversitÃ© Catholique de Louvain.

Schellmann, M., Zhao, Z., Lin, H., Siohan, P., Rajatheva, N., Luecken, V., and Ishaque, A. (2014). FBMC-based Air Interface for 5G Mobile: Challenges and Proposed Solutions. *9th International Conference on Cognitive Radio Oriented Wireless Networks and Communications* (CROWNCOM): 102-107, IEEE. https://doi.org/10.4108/icst.crowncom.2014.255708.

Stefanatos, S. Foukalas, F. and Khattab, T. (2017). On the Achievable Rates of OFDM with Common Phase Error Compensation in Phase Noise Channels. *IEEE Transactions on Communications*, 65(8): 3509-3521. https://doi.org/10.1109/TCOMM.2017.2705071.

Takeo, S., Asuka, M., Masanori, N., Seiji, O., Fukutaro, H., and Yoshiaki, K. (2019). Laser Phase Noise Tolerance of Uniform and Probabilistically Shaped QAM Signals for High Spectral Efficiency Systems. *IEEE/OSA Journal of Lightwave Technology*, 38(2): 439–446. https://doi.org/10.1109/JLT.2019.2945470.

Tao, J., Wu, J., and Xiao, C. (2009). Estimation of Channel Transfer Function and Carrier Frequency Offset for OFDM Systems with Phase Noise. *IEEE Transactions on Vehicular Technology*, 58(8): 4380-4387. https://doi.org/10.1109/TVT.2009.2020066.

Trung, N. Christophe, P. (2017). Kalman Filtering For Carrier Phase Recovery in Optical Offset-QAM Nyquist WDM Systems. *IEEE Photon Technology Letters*, 29(12): 1019-1022. https://doi.org/10.1109/LPT.2017.2701907.

Trung, N. François, R. Simon, G. Jérome, L. François, H. (2017). Efficient Chromatic Dispersion Compensation and Carrier Phase Tracking for Optical Fiber FBMC/OQAM Systems. *Journal of Lightwave Technology*, vol. 35(14): 2909-2916. https://doi.org/10.1109/JLT.2017.2707179.

Wu, B. Lu, J. Gao, M. Ren, H. Le, Z. Qin, Y. and Hu, W. (2020). Time-Domain Blind ICI Compensation in Coherent Optical FBMC/OQAM System. *Sensors*, 20 (21): 6397. https://doi.org/10.3390/s20216397.

Wu, B. Lu, J. Ren, H. Le, Z. and Hu, W. (2019). Time Domain Phase Noise Compensation in Coherent Optical FBMC/OQAM System. In *Proceedings of Asia Communications and Photonics Conference* (ACP 2019) (1-3). IEEE.

Xavier, A. M. (2015). *Prototyping FBMC/OQAM for 5G Mobile Communications Proof of Concept*. Master thesis, University of Surrey, UK. http://hdl.handle.net/2099.1/26604.

Xinwei, D. Song, T. Li, Y. Wu, M. W., and Kam, P. Y. (2020). An Optimum Signal Detection Approach to the Joint ML Estimation of Timing Offset, Carrier Frequency, and Phase Offset for Coherent Optical OFDM. *Journal of Lightwave Technology*, 39(6): 1629-1644. https://doi.org/10.1109/JLT.2020.3042546.

Xu, T., Liga, G., Lavery, D., Thomsen, B. C., Savory, S. J., Killey, R. I., and Bayvel, P. (2015). Equalization Enhanced Phase Noise in Nyquist-Spaced Super Channel Transmission Systems Using Multi-Channel Digital Back Propagation. *Scientific Reports*, 5(1): 1-13. https://doi.org/10.1038/srep13990.

Zhang, P., Liu, J., Shen, Y., and Jiang, X. (2021). Exploiting Channel Gain and Phase Noise for PHY-Layer Authentication in Massive MIMO Systems. *IEEE Transactions on Information Forensics and Security*, 16: 4265-4279. https://doi.org/10.1109/TIFS.2020.3029894.

Chapter 5

Indoor Environments' Thermal Comfort and Mitigation of Airborne Transmission of Viruses

Elaf S. Barrak[1,*]
Hasanain M. Hussain[2]
and Laith J. Habeeb[3]
[1]Mechanical Engineering Department, University of Wasit, Wasit, Iraq
[2]Mechanical Engineering Department, University of Technology– Iraq, Baghdad, Iraq
[3]Training and Workshop Center, University of Technology– Iraq, Baghdad, Iraq

Abstract

Human health and the thermal comfort of the individual have been of essential concern, especially in recent years. During the last COVID-19 pandemic, the spread of the disease became very aggressive, and human health was threatened. Significant health concerns from viral airborne transmission in enclosed spaces must be mitigated with efficient engineering control systems. The important features of this chapter are comprehensive coverage of indoor airborne contamination, practical insights, recommendations for mitigation strategies, and suggestions for future research directions. It aims to identify effective strategies and interventions to mitigate contamination, improve indoor air quality, and promote occupant well-being. The numerous engineering measures used to lessen viral transmission in indoor air conditions are thoroughly reviewed in this chapter, such as negative pressure methods, filtration systems, ventilation tactics, and air purification technology. The review reveals that viral particles are diluted and eliminated by adequate ventilation. It could be accomplished by enhanced outside air intake and

* Corresponding Author's Email: ebarrak@uowasit.edu.iq.

In: Advancements in Sustainability Systems
Editors: Hussain H. Al-Kayiem, Hasanain A. Abdul Wahhab and Raed A, Jessam
ISBN: 979-8-89113-571-0

better air dispersion. High-efficiency particles and other high-efficiency air (HEPA) filters capture and remove viruses and other pollutants. Bipolar ionization and ultraviolet germicidal irradiation are two air purification technologies that provide alternative ways to inactivate or eliminate viruses from the air. Particularly in healthcare institutions, negative pressure methods limit and trap infectious particles, reducing the danger of transmission. These engineering measures reduce viral propagation, but their efficiency relies on correct design, upkeep, and adherence to rules. These steps may make indoor spaces safer and reduce the spread of airborne viruses when used in conjunction with continuing investigations and controlled practices.

Keywords: airborne contaminant, COVID-19, indoor air quality, negative pressure, thermal comfort, ventilation

Abbreviations

ASHRAE	American Society of Heating, Refrigerating and Air-Conditioning.
CFD	Computational fluid dynamic.
CRE	Contaminant removal effectiveness.
HEPA	High-Efficiency Particulate Air.
HVAC	Heating, ventilating, and air conditioning.
IAQ	Indoor Air Quality.
IEQ	Indoor Environmental Quality.
PMV	Predicted mean vote.
PPD	Predicted percentage of dissatisfaction.
POV	Protected occupied ventilation
UV	Ultraviolet
mPMV	Modified Predicted Mean Vote.
MV	Mixed ventilation

Nomenclature

ε	Contaminant removal effectiveness.
C_e	Contaminant concentration at the exhaust.
C_s	Concentration of contaminants at the source.
C_b	Contaminant concentration at the typical contaminant.

C_{ex}	Contaminant concentration in the exhaled flow.
f_{cl}	Clothing surface area factor.
h_{conv}	Convection Heat Transfer Coefficient ($W/m^2 \cdot K$).
I_{cl}	Clothing insulation ($m^2 \cdot K/W$).
M	Metabolic rate (W/m^2).
P_w	Partial pressure of water vapor (N/m^2).
P_{ws}	Water vapor pressure at saturation (N/m^2).
T_a	Air temperature (°C).
T_{cl}	Clothing surface temperature (°C).
T_{mr}	Mean radiant temperature (°C).
T_n	Predicated neutral temperature (°C).
T_o	Outdoor temperature (°C).
T_{sk}	Average skin temperature.
W	Effective mechanical power (W/m^2).

1. Introduction

The requirement for comfort nowadays is higher than in the past when it was easier to produce livable conditions because of the rapid growth of industrial technology. It is discovered that energy consumption greatly depends on the comfort needs of the inhabitants as well as the features of the thermo-physical properties of the building materials. Buildings in a tropical climate greatly depend on air-conditioning and mechanical ventilation (ACMV), design, and utility system performance to achieve a comfortable indoor thermal environment for the occupants (Al-Kayiem et al., 2023; Sarbu and Sebarchievici, 2013).

The relevance of addressing indoor airborne contamination in the present scenario is critical due to the COVID-19 pandemic, increased indoor time, energy efficiency measures, urbanization, air pollution, and the direct impact on the well-being of occupants.

1.1. Thermal Comfort

"Thermal comfort" refers to the emotional state of being satisfied with one's thermal surroundings. Various factors influence it, including temperature, humidity, air movement, and personal preferences. Traditionally, heating and cooling systems in buildings have relied on fossil fuel-based energy sources, which contribute to greenhouse gas emissions and climate change. The growth of population and urbanization is often associated with an upward trend in energy demand and natural resources, such as fossil fuels, which burden the ecosystem (Al-Kayiem et al., 2020; Al-Kayiem et al., 2023; Yuan et al., 2022).

Renewable energy sources, such as solar and geothermal energy, offer alternatives that can contribute to sustainable and energy-efficient thermal comfort solutions. The development of knowledge and related technical developments over the last few centuries has led to an evolution of human civilizations overlying nature, endangering natural systems. The need to promote sustainable development, as articulated in the Brunt Land Report (WCED, 1987), stems from a growing awareness of environmental contamination and the depletion of natural resources (Commission on Environment, 1987).

1.2. Indoor Airborne Contamination

Typically, people spend between 80-90% of their time inside. Employee costs account for most of the company's running costs, with just 1-3% related to energy costs and 8 to 11 percent to other facility expenditures (Gries & Schneider, n.d.). Since ASHRAE standards were published, there has been increased interest in both academic and practitioner literature on occupant health and building design. Studies have shown that various comfort- and health-related impacts are connected to the features of the structure (Thermal Environmental Conditions for Human Occupancy, 2004). Office building occupant performance has also received significant attention from scholars and practitioners (Bluyssen, 1996).

Regulating indoor air quality (IAQ) is reviewed by different terminologies. IAQ in workspaces is crucial to protecting workers from contamination due to indoor pollutants, as it directly affects their health and well-being (Al horr et al., 2016).

IAQ concerns, such as those involving the building's thermal, acoustic, visual, and air quality, directly influence its inhabitants' comfort, health, and productivity (De Giuli et al., 2012). With the recent outbreak of COVID-19 disease, there has been much research done on the droplets of breathing, sneezing, and coughing. Previous literature used two methods for this research area: an experiment and numerical simulation by computational fluid dynamics (CFD) to follow the transmission of airborne virus infection from person to person in indoor and outdoor environments (Aljabair et al., 2023; Tang et al., 2021). Thus, based on the analysis, it has been noticed that factors of engineering technique in indoor design that help to eliminate airborne contaminants and enhance thermal comfort, which plays a vital role in overall occupant productivity, were presented in the present systematic review in Table 1.

Table 1. Summary of the paper review in the last decade

Author (year)	Location	Conclusion / Remarks
Gola et al., 2019	Italy	Rise to a broad scenario on the existing knowledge regarding indoor air pollution, design and management strategies for healthy spaces, and emerging topics.
Elsaid & Ahmed, 2021	Egypt	provides insight into the design of air-conditioners and the impact on the spread of the hybrid coronavirus with recommendations.
Fadaei, 2021	Iran	Review the importance of appropriate ventilation and how it can effectively prevent the expansion of COVID-19.
Noorimotlagh et al., 2021	Iran	improve indoor air environments, such as improving buildings' ventilation indoors, and observe an interpersonal distance of more than 2 m.
Agarwal et al., 2021	India	review of all the improvement techniques bearing in mind thermal comfort and safety of occupants and offers recommendations for a sustainable future.
Piscitelli et al., 2022	Sri Lanka	It intends to outline the literature concerning the transmission of viruses in different environmental settings and study the virus's behavior to prevent infection.
Ding et al., 2022	Netherlands	discussed the possibility of changing natural or mixed mechanical ventilation to personalized ventilation in school classrooms to reduce airborne transmission.
Ferrari et al., 2022	Italy	Studied the ventilation strategies report that programmed mechanical ventilation systems can reduce risk-infection during pandemic events for school classrooms.

1.2.1. Background Overview of Indoor Air Quality

Several investigations used a breathing thermal manikin with realistic respiration simulations with different ventilation modes. They tried to choose the most suitable way to minimize the indoor pollution particles and keep the thermal comfort range for people in safe health in their buildings. Various laboratory and real-world studies have reported associations between ventilation rate and IAQ measures. These associations have been particularly observed in schools, hospitals, buildings, and offices.

Protected zone ventilation (PZEV) was experimentally tested by Cao et al., 2018. The findings suggest that even when two people are standing close to one another, shielded zone ventilation may drastically reduce personal exposure to the other person's exhaled airflow. Compared to displacement ventilation, exposure estimates from protected zone ventilation could be 20 times lower. Li et al., 2015 studied the POV system, which can deflect a high-motion cough jet to lessen the chance of exposure to someone else's coughed-up particles.

Morawska, 2006 discussed the importance of deepening knowledge and developing an understanding of the physics and microbiology of virus transport and their survival in the air to limit or prevent viral spread. It also suggested that controlling and preventing the spread of viral diseases in healthcare settings, public places, and indoor and outdoor environments requires a greater awareness of how viruses propagate. Therefore, one strategy to stop the transmission of airborne viruses inside buildings may be to improve ventilation systems. Chow et al., 2006a proved numerically that the Negative pressure in an operating theater can decrease the spread of airborne infections by creating a pressure gradient that prevents contaminated air from escaping the room and entering other areas. In a negative-pressure environment, the air is constantly being drawn into the room and then exhausted outside, which helps to remove any infectious particles that may be present. This reduces the risk of infection exposure for patients and healthcare personnel outside the operation room. Further lowering the danger of infection transmission by adopting a strong low-level exhaust system without flow recirculation could make it possible for the air to exchange quickly.

Cao et al., 2015 tested two types of ventilation systems, PZV and HPZV, in an office room with two breathing manikins to see if they could reduce air transmission between people and, therefore, the risk of airborne cross-infection. The results suggest that these systems may effectively reduce exposure to indoor pollutants.

How a person walking near someone using personalized ventilation (PV) affects the quality of the air being breathed in was studied by (Al Assaad et al., 2019. They found that when the walking person was too close, it decreased the efficiency of the PV and increased the number of contaminants in the air, but using a wider PV helped. Wang et al., 2020 discussed using a vortex ventilation system to effectively remove airborne pollutants from indoor working zones, including viruses. The vortex ventilation system forms an "indoor tornado" that quickly inhales pollutants and extracts them through a vortex tube, preventing unorganized diffusion of the pollutants and reducing their mean residence time in the working zone and suggesting that the location of the pollution source is important for the effectiveness of the vortex ventilation system and that pollutants released near the vortex core are significantly affected by the negative pressure gradient of the vortex ventilation, limiting their diffusion in the occupied zone. Therefore, using a vortex ventilation system can potentially eliminate the spread of airborne transmission of viruses in indoor air environments.

Measurement systems are vital in assessing indoor airborne contaminants' concentration levels and characteristics. They enable the collection of real-time or periodic data regarding pollutant levels, particle sizes, chemical compositions, and microbial presence. These measurements provide valuable insights into the effectiveness of mitigation strategies, the impact of ventilation systems, and the evaluation of indoor air quality standards.

The development and utilization of modern measurement systems with high accuracy have significantly contributed to indoor air quality research advancements. Advanced techniques such as real-time monitoring, remote sensing, and advanced sampling methods have enhanced our understanding of indoor airborne contaminants' temporal and spatial variations. Moreover, integrating measurement systems with numerical analysis and modeling approaches can facilitate the validation and calibration of simulation models, ensuring their accuracy and reliability. The combination of measurements and modeling allows for a more comprehensive assessment of indoor air quality and aids in identifying potential sources and pathways of contamination.

Future research should continue to focus on developing and refining measurement systems for indoor airborne contamination. This includes advancements in sensor technologies, data acquisition methods, and data analysis techniques. Additionally, efforts should be made to standardize measurement protocols and establish guidelines for accurate and consistent data collection. By addressing the importance of measurement systems and their impact on characterizing indoor airborne contamination, researchers and practitioners can make informed decisions, improve the accuracy of numerical analysis, and effectively implement strategies to mitigate indoor air pollution and enhance occupant health and well-being.

2. Engineering Controls

Some factors can eliminate the spread of airborne transmission of viruses in the indoor air environment, improve indoor air quality, and help to mitigate any future bio-attack. Several studies covered the study of biology, chemistry, engineering, information, and physical and human resources, especially during the COVID-19 pandemic. The important information was extracted from recent publications journals from Google Scholar, ScienceDirect, Scopus, Wiley, SpringerLink, and PubMed. Figure 1 illustrates the traditional ways to minimize airborne transmission in indoor air quality by engineering controls.

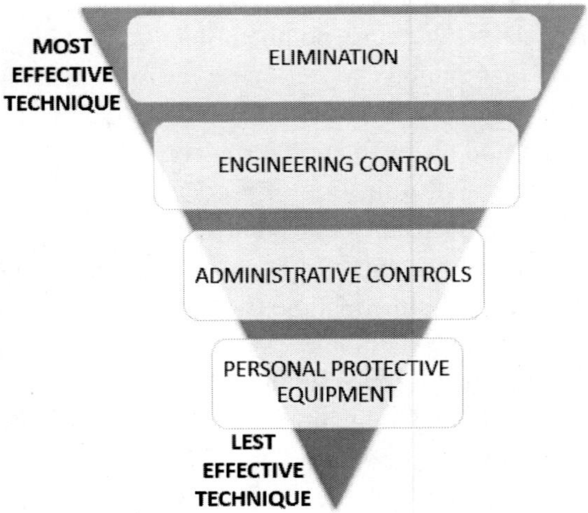

Source: Authors.

Figure 1. The US Centers for Disease Control's conventional infection control pyramid.

To reduce the spread of IAQ-engineered measures aimed at reducing the possibility of airborne transmission by Ventilation, Filtration, Air Distribution, Disinfection, and Temperature and Humidity Controls. Ventilation is crucial; it "refers to supplying fresh air to an indoor ambient and exhausting polluted air." Two fundamental components of a room's ventilation strategy are air distribution and ventilation rate, which may be achieved naturally, artificially, mechanically, or both (hybrid).

The ventilation rate is often given as L/s (m^3/h) per person or L/s (m^3/h) per m^2 of floor space (Ding et al., 2022). Usually, hospitals are naturally ventilated, including several critical care rooms. Yet, suppose the airflow pathway is blocked (for example, by shutting windows and doors). In that case, the concentration of airborne pathogens may quickly increase, increasing the danger of airborne infection and transmission (Gilkeson et al., 2013; Adam et al., 2021). Good ventilation techniques are already in place in many medical settings daily, and emergency precautions against droplet and contact transmission are illustrated in the design guide proposed by Health Building Note 00-01, 2014, outlined in Figure 2. The passengers are also shielded against airborne transmission by adequate ventilation. When necessary (like during the COVID-19 pandemic), the ability to boost ventilation rates may vary and may be constrained in some way by their initial design requirements and execution. There are several types of ventilation, including mixing ventilation (MV), displacement ventilation (DV), personalized ventilation (PV), hybrid air distribution (HAD), stratum ventilation (SV), protected occupied zone ventilation (POV), local exhaust ventilation (LEV), and piston ventilation (PiV).

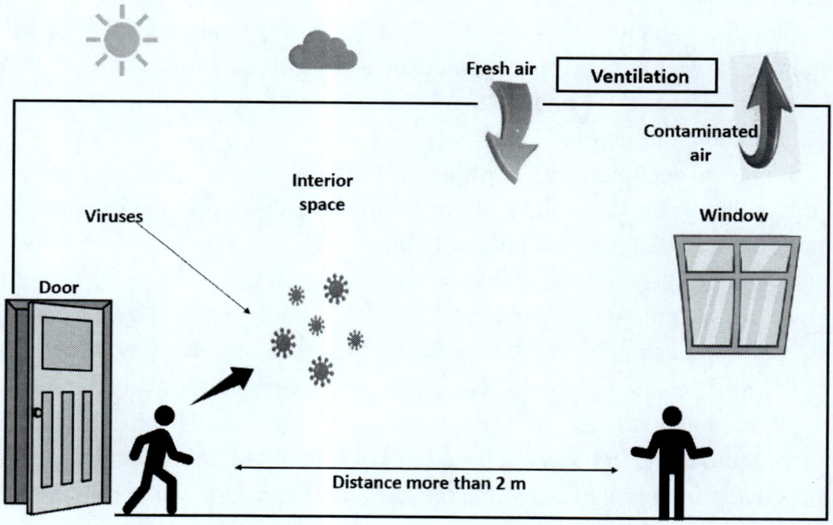

Source: Authors.

Figure 2. Engineering level controls to reduce the environmental risks for airborne transmission.

Protected zone ventilation may be helpful in certain circumstances, such as in hospitals and healthcare institutions, to prevent cross-infection based on direct and indirect exposure to infectious microorganisms created by humans. Liu et al., 2022 studied the POV system strategy. The POV can deflect a high-motion cough jet to lessen the chance of exposure to someone else's coughed-up particles. Compared to an MV system, the receiver occupant's peak exposure level and the intake fraction of coughed particles are down an order of magnitude. According to the findings, the normalized average peak concentration in the respiratory zone for MV is 16.2%. Two-slot POV is 3.18%, and for one-slot POV, it is 1.78%. Comparing the one-slot POV to using a regular MV, the peak concentration is reduced by an order of magnitude.

Cao et al., 2015 experimented to evaluate the effectiveness of protected zone ventilation (PZEV) and hybrid protected zone ventilation (HPZV) in minimizing exposure to exhaled air from other people's respiration. Consequently, the risk of respiratory cross-infection between two individuals in similar relative positions would probably be reduced. The experimental results demonstrated that in both the PZV and the HPZV systems, it is possible to reduce the transmission of tracer gas from one manikin to the opposite manikin. The results suggest that even when two people are standing close to one another, shielded zone ventilation may drastically reduce personal exposure to the other person's exhaled airflow. Protected zone ventilation may result in exposure estimates that are 20 times lower than those from displacement ventilation.

Protected occupied zone ventilation (POV) was tested in hospital isolation wards to lower the possibility of cross-infection. A sitting receiving medical practitioner and an infected patient lying on a bed was portrayed by two life-size breathing thermal manikins by Aganovic et al., 2019. To replicate the droplet nuclei produced by patients, N_2O was employed as a tracer gas. The outcome demonstrated the value of the ventilation approach. It was found that the slot diffuser's supply velocity significantly impacted how much pollution the HCW was exposed to. The proximity of the exhaust pipe to the sick patient's breathing zone had a little but considerable impact on the exposure index. The findings also show that the danger of exposure while utilizing POV is much lower than when using the often-used MV in hospital wards.

Aganovic et al., 2019 also estimated the proportion of individuals dissatisfied with the airflow near the healthcare worker (HCW) and patient in a single-bed isolation room equipped with a positive-pressure ventilation system. According to the CFD validation results for the three examined turbulence models, the SST k-ω models correctly foretold the velocity and temperature profiles based on the experimental results. The SST k-model was also used to analyze the effects of supply velocity and temperature on drought risk and to illustrate the ventilation distribution in the patient ward. Even at reduced supply temperatures, supply velocities of 1.0, 1.5, and 2.0 m/s were not higher than the suggested comfort threshold of 20%. The risk of ventilation at the ankle level of the seated patient exceeded the upper limit permitted for supply velocities of 2.5 and 3.0 m/s at both lower and higher supply temperatures.

Berlanga et al., 2018 investigated the implementation of the DV strategy in AIIR as an alternative to the most extensive MV approach using a representative case study of an AIIR setup. A common AIIR configuration involving two thermal breathing mannequins and a radiant wall modeling an exterior wall considers the thermal comfort of the chamber's occupants, ventilation, and contaminant performance indices, as well as the contact of the health worker (HW) to the pollutants breathed by the confined patient (P). Three air renewal rates are examined to determine whether they affect the variables under study. The results indicate that both manikins' ventilation, pollutant, and overall comfort indices perform well in the cases examined.

Alashaab and Alamery, 2018 investigated ways to make mosques in Iraq more comfortable thermally. Baghdad's Omar bin Abdul Aziz Mosque is used as a case study. Four ventilation fans provided air ventilation in the mosque to lower the CO_2 content at the breathing level. Getting the values of the expected mean vote allowed for thermal comfort (PMV) evaluation. When four fans are added for additional ventilation, PPD increases by 10%, PMV increases by 14%, and the concentration of carbon dioxide at the breathing level increases by 0.0015 kg/kg_{air}. The results demonstrate that the three thermal comfort levels should be taken for the surgery room's air conditioning calculations.

Other investigations concentrated on extracting indoor airborne contaminants using negative pressure methods. (Chow et al., 2006b) address the above topics, including the physical conversion of a conventional positive-pressure operating room into a negative-pressure environment, problems encountered, airflow design, and performance evaluation. Computational fluid dynamics, a type of computer modeling, was used to examine the distribution of chamber air before and after the design modifications from positive to negative pressure. According to simulation results, the physical environment and bacterial dispersion pattern in the negative pressure theater were at least as good as in the original positive pressure design. In this instance, ongoing testing and air sample tests confirmed the quality of the finished product. In addition, the steady-state modeling revealed that the bacterial dispersion behavior and thermal comfort of the negative pressure theater were comparable to or even superior to the original positive pressure design.

Two negative-pressure operating theater simulation models were used in a computer analysis by Chow et al., 2006a. The commercial software FLUENT studied the airflow behavior and dispersion of infectious particles discharged from surgical personnel and patients in a specially built negative-pressure operating theater. The analyses showed that the negative pressure operating room airflow performance was good and comparable to the original positive pressure design. The airflow design proved effective in halting the transmission of infectious particles. According to their calculations, the airflow limited the patient's discharge of infectious particles far enough to protect the working environment for the medical and surgical teams. The simulation outputs demonstrated that the ventilation system in our operating theater with negative pressure accomplished the desired results by protecting the patient and the medical team from the airborne virus.

Cheong & Phua, 2006 investigated the airflow and pollutant dispersion patterns in a "negative pressure" isolation chamber using objective measurement and CFD modeling. As part of the initial investigation utilizing Strategy 1, two air supply diffusers and two air exhaust grilles were installed on the ceiling of the isolation chamber.

The two extract grilles are moved to the wall behind the bed, at a height of 0.3 meters above the floor, while the air supply diffusers from Strategy 1 remain in position. In Strategy 3, which has the same layout as Strategy 2, the supply grilles are moved closer to the wall behind the bed in place of ceiling diffusers. The supply air grilles provide occupants with a laminar flow of outside air while minimizing the entrainment of interior air, making Strategy 3 the most effective at removing pollutants. The airflow and pollutant dispersion patterns in the isolation room are found to be significantly affected by the ventilation strategy and furniture configuration.

To evaluate the containment and removal of respirable aerosols (0.5-10 m) in a general patient room and an airborne infectious isolation room regarding ventilation rate and directional airflow, Mousavi et al., 2014 performed a series of studies in a real hospital. The suspension and migration of larger particles (3-10 m) throughout the patient room(s) and the particle concentrations along the routes between the source and the exhaust increased by more than 40%. This study suggests that negative pressure, as recommended by healthcare ventilation standards such as ASHRAE 170-2008 (ASHRAE, 2008), facilitates the limitation of aerosol movement. However, the results suggest that increased air change rates may not be as effective as previously thought in removing infectious aerosols from patient rooms and may even increase exposure to particulates suspended in turbulent airflow in the respiratory zone.

Offermann et al., 2016a carried out a one-week experiment. Aerosol was released in a patient's room with and without modifications to the ventilation system for surge control, and concentrations in the room and corridor were measured. The average corridor protection efficiencies were high (>98%) in the baseline ventilation configuration and when the ventilation system was modified for complete negative pressure on the floor, indicating that very little PM entered the corridor. During the trial, a tracer aerosol (polyalphaolefin, PAO) was released at bed height in the patient chamber to mimic the constant production of airborne particles. The baseline and modified ventilation arrangements demonstrated outstanding protection efficiencies, which decreased when the door opened. For maximal infection control and to limit the migration of particles into the hallway, the doors should be kept shut as much as possible. In addition, if transmission of airborne infections is a concern, an anteroom can be utilized to reduce particulate transfer from patient rooms further toward hallways.

Table 2. Summary of previous studies on different mitigating techniques used

Author (Year)	Methodology	indoors	Techniques
Karottki et al., 2013	Experimental	Homes	Air purifying respirators with or without HEPA filtration.
Chuang et al., 2013	Experimental	Cars	AC operation modes.
Berlanga et al., 2018	Experimental	AIIR	Displacement ventilation strategy.
Al Assaad et al., 2018	Simulation	Office	Indoor air disturbance with the conventional Mixing Ventilation system.
Cao et al., 2018	Experimental	Full-scale office	increased air distribution in the operating zone
Cheng et al., 2021	Experimental & simulation	Small climate chamber	Localized laminar airflow ventilation with mixing ventilation.
Zhang et al., 2022	Simulation	Full-scale room.	Negative pressure room technique.
Alkhalaf et al., 2023	Simulation	Hospitals	Various ventilation strategies were designed.

Two patients and one healthcare provider were in a two-bed hospital ward utilizing stratum ventilation, which incorporated displacement ventilation. Lu et al., 2020 and Mousavi and Grosskopf, 2014 utilized a numerical model to examine pollutant dispersion. CO_2 was used as a gas tracer representing the contaminants that patients vomited and inhaled. Two patients inhaling under stratum ventilation resulted in a decrease in pollutant concentration and an increase in contaminant removal efficiency in the respiratory zone (1.3–1.7 m above the floor). Comparing the contaminant concentration at various phases following a patient's cough reveals that the aspirated contamination is swiftly diluted by stratum ventilation, and the high concentration site is significantly diminished. Results indicate that stratification ventilation reduces the risk of exposure for hospital ward personnel. Related studies performed in this field in the last decade are summarized in Table 2.

3. Mathematical Methods of Engineering Controls

A few pertinent principles, including mathematical modeling and computations, are proposed to identify engineering controls for minimizing airborne transmission of viruses in indoor environments.

1. Calculations for Airflow and Ventilation: Based on variables like room size, occupancy, and pollutant concentration, engineers often employ mathematical equations to calculate the necessary airflow rates, air change rates, or ventilation rates for areas. These calculations make sure that airborne pollutants are sufficiently diluted and removed.
2. Calculations of Filtration Efficiency: Mathematical formulas may be used to determine the effectiveness of air filters, especially HEPA filters. The filter's capacity to catch particles of a certain size is measured using established tests, such as the Minimum Efficiency Reporting Value (MERV) rating.
3. Contaminant Transport Modeling: Mathematical simulations of the movement and dispersion of airborne pollutants inside enclosed spaces, including CFD. These models include fluid flow, turbulence, and particle transport equations to forecast contaminant behavior and assess the efficacy of engineering control measures.
4. Infection Risk Assessment: Using variables like the concentration of infectious aerosols, ventilation rates, and occupancy, mathematical models like the Wells-Riley equation are used to calculate the risk of infection in indoor settings.

These examples show how mathematical equations are used in several engineering controls and components of indoor air quality. Still, it's crucial to note that the precise equations and computations might change based on the conditions, particular objectives, and accessible data. More thorough mathematical formulations and models tailored to circumstances may be found by consulting pertinent research papers and publications in indoor air quality and engineering controls. In this chapter, two main important factors affect the performance of the airborne contaminants' removal efficiency and the design of the indoor airflow for thermal comfort by using appropriate ventilation system strategies:

3.1. Predictions of Contaminate Removal

The system's effectiveness in removing pollutants from the breathing zone is assessed using a metric called Contaminant Removal Efficacy (CRE) (Cao et al., 2014; Lu et al., 2020). The pollutant sources' location, characteristics, and airflow patterns affect the CRE (Liu et al., 2022). Contaminant removal performance decreases with decreasing contaminant size (Morawska, 2006). CRE is calculated using Equation 1.

$$\text{CRF} = \frac{C_e - C_s}{C_b - C_s} \quad (1)$$

The terms C_e, C_s, and C_b represent the contaminant concentrations at the exhaust, supply, and average contaminant concentrations in the respiratory zone, at 1.3-1.7 m above the floor. The ratio of contaminant concentration, C, to source concentration in a volume is expressed using the relative contaminant concentration, which is computed using equation 2:

$$C = \frac{C_b}{C_{ex}} \quad (2)$$

The number of contaminants present in the exhaled flow is known as C_{ex}.

3.2. Prediction of Thermal Comfort

3.2.1. Fanger's PMV Equation

The predicated mean vote (PMV) relies on the heat stability of the human body, in which the body's internal heat emission plays a significant role, and heat loss to the environment is equal. The PMV method combines two personal variables, activity level, and clothing insulation, with four physical variables: air temperature, air velocity, mean radiant temperature, and relative humidity. The PMV was developed as an index that can be used to predict the average thermal experience of a large population in space. The PMV model and definitions of its parameters are presented in the set of Equations 3 to 7.

$$\text{PMV} = \left(0.303\, e^{(-0.036 \cdot M)} + 0.028\right)\{ (M-W) - 3.05\,[5.733 - 0.007(M-W) - 0.001P_w] - 0.42[(M-W) - 58.15] - 0.0173M\,(5.867 - 0.001P_w) - 0.0014M\,(34 - t_a) - 3.96 * 10^{-8} f_{cl}\,[(t_{cl}+273)^4 - (t_{MR}+273)^4] - f_{cl} \cdot h_c \cdot (t_{cl} - t_a)\} \quad (3)$$

$$t_{cl} = 35.7 - 0.0275(M-W) - R_{cl}\{3.96 * 10^{-8} f_{cl}\,[(t_{cl}+273)^4\,] - f_{cl} \cdot h_c \cdot (t_{cl} - t_a)\} \quad (4)$$

$$f_{cl} = 1.0 + 0.2I_{cl} \quad (5)$$

where,

$I_{cl} < 0.5clo$ $1.05 + 0.1I_{cl}$, and $I_{cl} > 0.5clo$

$$h_c = 2.38(t_{cl} - t_a)^{0.25} \quad (6)$$

where,

$2.38(t_{cl} - t_a)^{0.25} > 12.1\sqrt{V}\; 12.1\sqrt{V}$, *and*
$2.38(t_{cl} - t_a)^{0.25} < 12.1\sqrt{V}$

$$R_{cl} = 0.155I_{cl} \quad (7)$$

3.2.2. The Modified PMV Model

The modified predicated mean vote (mPMV) is an adaptation of the PMV model, widely used to assess thermal comfort in indoor environments. The mPMV model incorporates additional factors and considerations to improve the accuracy of thermal comfort predictions.

Temperature, humidity, air velocity, and clothing insulation. It considers occupants' physiological responses and subjective perceptions to estimate thermal comfort levels (Al-Kayiem et al., 2023).

Modifying the mPMV model involves refining the mathematical equations and adjusting the coefficients used in the PMV model. These modifications are often based on empirical studies and field measurements to enhance the accuracy and applicability of the model across a range of indoor environments and occupant conditions (Laouadi, 2022).

By incorporating the additional factors and adjustments, the mPMV model, given in equation 8, aims to provide more accurate thermal comfort predictions and better align with actual occupant experiences. It enables designers, engineers, and facility managers to optimize indoor environmental conditions and HVAC system settings to enhance occupant comfort and satisfaction. The mPMV model continues to be refined and updated as new research and insights emerge in thermal comfort. It is a valuable tool in designing and evaluating indoor environments, facilitating the creation of comfortable and energy-efficient spaces for occupants.

$$mPMV = [0.303\ e^{-0.036}(0.1092 * \exp^{(MAP * 0.0296)} + 0.028]\ \{[(0.1092 * exp^{(MAP * 0.0296)}) - W] - 3.05 * 10^{-3}\ [5733 - 6.99\ ([(0.1092 * exp^{(MAP * 0.0296)}) - W) - p_a] - 0.42[(0.1092 * exp^{(MAP * 0.0296)}) - W) - 58.15] - 1.7 * 10^{5} * (0.1092 * exp^{(MAP * 0.0296)})\ (58.15 - p_a) - 0.0014(0.1092 * exp^{(MAP * 0.0296)})\ (34 - T_a) - 3.96 * 10^{-8}\ fcl\ [(T_{cl} + 273)^4 - (T_{mrt} + 273)^4] - f_{cl}\ h_c\ (T_{cl} - T_a)\} \quad (8)$$

Recently, Al-Kayiem et al., 2023 contributed to the PMV by modifying the related equations to satisfy the weather conditions in tropical areas. The study found that the mPMV, equation 13, showed that the participants are likely to feel comfortable when the indoor temperature is between 24°C and 26 °C. The PMV model overestimated the thermal sensation in the room. Accordingly, they developed a new indicator for thermal comfort named the maPMV model to mitigate the inaccuracy of the existing PMV model, equation 3, and mPMV model, equation 13, in tropical and hot weather conditions. The coefficients of the developed model were determined based on actual thermal sensation data obtained from a questionnaire survey of the occupants in two experimental cases shown in Table 3.

Table 3. Cases of Al-Kayiem et al., 2023 study to develop the modified adaptive predictive mean vote in tropical and hot weather conditions [with permission]

case	# of participants	Statistical prediction	Metabolic rate (met)	Clothing insulation (clo)
Case 1	10	Average	1.02	0.563
		Standard deviation	0.0632	0.0718
Case 2	20	Average	1.3	0.588
		Standard deviation	0.0165	0.0544

Equations 9a and 9b represent the newly developed maPMV model for estimating tropical and hot weather thermal comfort.

$$\text{Case 1, } maPMV = \frac{PMV}{1+\lambda*PMV}, \qquad \lambda = 0.2953 \tag{9a}$$

$$\text{Case 2, } maPMV = \frac{PMV}{1+\lambda*PMV}, \qquad \lambda = 2.0506 \tag{9b}$$

where PMV is given by Equation 3.

4. Discussion

Due to the importance of occupants' health and thermal comfort, the World Health Organization and its emphasis on thermal comfort focused on offering clean air, which impacts the regular metabolism of our body. In total, 41 articles were eligible for a full evaluation. The association between HVAC systems in hospitals, schools, restaurants, and buildings was investigated in ten (10) experimental articles using mannequins for patients, nurses, and doctors in hospitals and students in schools, as well as tracer gases like CO2, NOx, dust, and other gases to track the pathogens of airborne contaminants. Nineteen (19) publications utilized simulation because it was easy, cheaper, and gave more time to remove dispersed airborne pollutants. The remaining articles combined simulation and experimentation.

Figure 3 shows the specification type of the design article. The current ventilation tactics and numerous techniques to stop the spread of contaminants and get rid of viral infections like the flu, COVID-19, and other infectious diseases were examined in this systematic study.

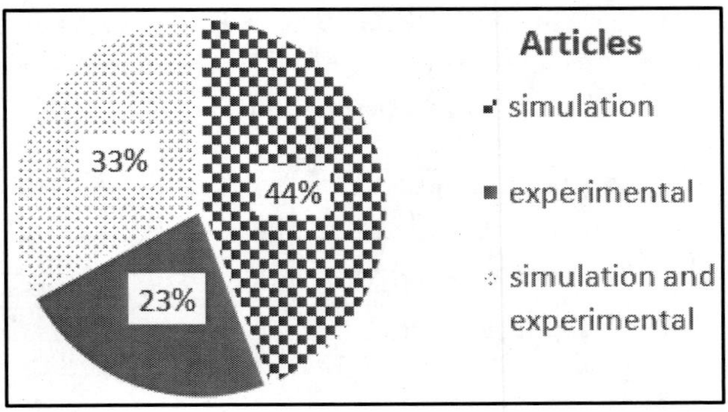

Figure 3. Analysis methods of methods articles.

Several strategies can be employed to establish a protected occupied zone with effective ventilation to reduce the transmission of viruses:

1. Local Exhaust Ventilation: Installing local exhaust ventilation systems close to probable viral particle sources, such as coughing or sneezing, may help to catch and eliminate these particles in the surrounding area of the source.
2. High-Efficiency Air Filtration: Insert a HEPA filter into the ventilation system, which effectively captures many viruses and other airborne particles.
3. Increased Outdoor Air Supply: To reduce the concentration of interior toxins and increase the ventilation system's intake of clean outdoor air. Bringing in clean air assists in lowering the total presence of virus particles.
4. Improved Air Distribution: Optimize the air distribution throughout the area to maintain optimum mixing and circulation. This facilitates the consistent distribution of fresh air throughout the inhabited area and lowers the concentration of airborne pollutants.
5. Designating negative pressure rooms or isolation rooms might help stop the transmission of infections in certain situations, such as healthcare institutions. To prevent infected air from escaping the room and possibly infecting others, these rooms must include ventilation systems that maintain lower air pressure than the surrounding environments, as described in Figure 4.
6. UV Germicidal Irradiation: To inactivate or eradicate viruses and other germs, ultraviolet (UV) germicidal irradiation devices may be placed inside the ventilation system. By disinfecting the air traveling through the system with UV light, these systems lower the danger of viral transmission.

It's crucial to remember that while effective ventilation techniques can help stop the spread of viruses, they should be used with other preventive measures like immunization, physical seclusion, mask use, and hand hygiene to offer complete immunity from viral transmission.

Additionally, depending on the size and function of the inhabited zone, different ventilation needs may apply; thus, speaking with ventilation experts or specialists is advised for customized solutions. The main goals of the discussion section are the analysis and interpretation of the data given in the article overview of engineering measures for reducing viral airborne transmission in indoor settings. It offers the chance to look more deeply into the evaluated engineering control measures' merits, constraints, and ramifications.

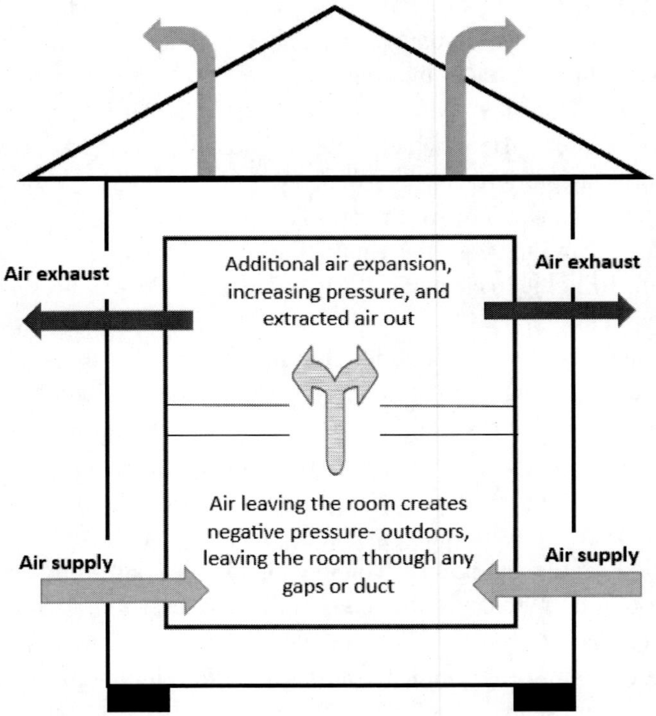

Figure 4. Negative pressure strategies.

4.1. Effectiveness of Engineering Controls

The reviewed engineering controls have demonstrated significant effectiveness in reducing the transmission of airborne viruses in indoor environments. Viral particles may be successfully diluted and removed using ventilation techniques, including enhanced external air supply and local exhaust ventilation. Improved interior air quality and reduced virus transmission benefit from properly built and maintained ventilation systems. Filtration systems have effectively captured and removed viral pollutants from the air, especially those with high-efficiency air filters like HEPA filters. It has been shown that applying air purification technologies, such as bipolar ionization and UV germicidal irradiation, may inactivate or eliminate viruses. These measures offer an additional degree of defense against airborne transmission.

4.2. Limitations and Challenges

Engineering controls have proven useful, but it's crucial to recognize their drawbacks and difficulties. These measures' performance strongly depends on their correct design, implementation, and upkeep. Poor system design, installation, or maintenance may affect the performance of ventilation, filtration, and air purification systems. Furthermore, engineering restrictions may or may not be successful depending on the viral strain and its transmission mode. It is essential to consider both the infectious dosage necessary for transmission and the size and dispersion of virus particles. Adopting and maintaining these controls under the related budgetary and logistical restrictions may be difficult, particularly in contexts with limited resources.

4.3. Implications and Recommendations

Effective engineering controls may greatly lower the risk of viral transmission in interior settings and other preventative measures, including immunization, physical separation, mask use, and hand cleanliness. To reduce the transmission of infectious illnesses, healthcare institutions should emphasize using negative pressure methods, isolation rooms, and improved ventilation systems. The results also emphasize the need for continuing study and improvements in engineering control technology. More studies may examine how to optimize ventilation rates, filtration efficiency, and air purification methods to increase the effectiveness of these procedures. Guidelines and standards must also be created to guarantee correct engineering control design, installation, and maintenance in various contexts.

4.4. Summary

In the present paper, indoor airborne contamination significantly contributes to the knowledge base by providing a comprehensive understanding of the engineering control strategies and their effect on the health implications of indoor airborne contaminants. It explores various strategies and interventions to mitigate contamination, improve indoor air quality, and promote occupant well-being. Its comprehensive approach and comprehensive subject matter analysis set it apart from other related reviews. It incorporates the latest research findings, industry best practices, and real-world examples to offer practical insights for addressing indoor airborne contamination. Furthermore, this review examines the impact of indoor airborne contamination on human health and highlights the importance of ventilation systems, air filtration, and proper cleaning practices in mitigating contamination. It provides valuable recommendations for improving indoor air quality and creating healthier indoor environments. The optimization in engineering techniques can involve several aspects, including:

1. Design Optimization: Assessing different design parameters or factors such as airflow patterns, ventilation system configurations, filtration efficiency, or placement of air outlets to determine the most effective design for reducing airborne transmission.
2. Performance Optimization: Evaluating the performance of engineering techniques under different occupancy scenarios, room layouts, and ventilation rates to identify the most effective measures for mitigating viral transmission.
3. Energy Efficiency Optimization: Balancing the need for effective viral transmission control with energy efficiency considerations, optimizing engineering techniques to achieve both goals.
4. Cost Optimization: Considering the cost-effectiveness of different engineering techniques and identifying approaches that best reduce viral transmission.

4.5. Perspective for the Future

While this chapter provides a solid foundation for understanding indoor airborne contamination, there are several avenues for future research and development. Some suggestions for future studies and advancements in this field include:

1. Conducting longitudinal studies to assess the long-term effects of exposure to indoor airborne contaminants on human health.

2. Investigation of emerging contaminants and their impact on indoor air quality, such as microplastics or chemicals from electronic devices.
3. Evaluation of the effectiveness of innovative technologies and interventions, such as smart ventilation systems, advanced air purification methods, and building materials with low emissions.
4. Assessing the impact of occupant behavior, lifestyle choices, and activity patterns on indoor airborne contamination levels
5. Development of indoor air quality management guidelines and standards, considering different building types, occupancy scenarios, and geographical locations.

By addressing knowledge gaps and implementing future suggestions, we can further enhance our understanding of indoor airborne contamination and develop more effective strategies to improve indoor air quality and safeguard occupant health.

This chapter significantly adds to the knowledge base by comprehensively understanding indoor airborne contamination. It stands out from other related reviews because of its comprehensive approach, practical recommendations, and coverage of various contaminants and mitigation strategies. The insights gained from this review can guide policymakers, researchers, and professionals in creating healthier indoor environments and improving the well-being of occupants.

Conclusion

The systematic review's objective was to gather and assess parameters already available in the literature on indoor thermal comfort and virus spread. These parameters are essential for preventing the spread of viruses via the air in enclosed spaces. Various engineering techniques were used to improve occupant safety while lowering the probability of viral propagation of viruses. Viral particles are successfully diluted and removed from interior areas by ventilation systems with high-efficiency air filters and an improved outside air supply. While air purification methods like UV germicidal irradiation and bipolar ionization help inactivate or remove viruses from the air, local exhaust ventilation aids in capturing and eliminating pollutants at their source. By putting these engineering control measures into place and adhering to public health recommendations, it is possible to make indoor spaces safer, reduce the spread of viruses in the air, and safeguard the inhabitants' health and well-being. Negative pressure engineering controls are also essential for preventing viruses and pollutants transmission in enclosed spaces. The air pressure in these regions is maintained lower than the surroundings by creating negative pressure rooms or isolation rooms. This design's excellent containment and confinement of potentially infectious particles decreases the danger of transmission to medical personnel and other patients. A controlled environment is created via negative pressure rooms, together with suitable ventilation and filtration, which reduces the escape of airborne pollutants and aids in maintaining a safe workplace. This review has successfully achieved its objectives by examining various aspects of indoor airborne contamination as follows:

1. To emphasize building managers, hospital administrators, and infection control teams the effectiveness of engineering controls in reducing the risks of airborne infection, particularly in the case of SARS-CoV-2.
2. To highlight the importance of increasing existing ventilation rates and improving ventilation effectiveness by utilizing current systems. This involves enhancing outdoor air change rates and eliminating air recirculation within the ventilation system to ensure a supply of fresh outdoor air.
3. To recommend using portable air cleaners with mechanical filtration systems in areas with known air stagnation or inadequate ventilation with the existing system. This is particularly relevant in locations such as influenza and COVID-19 cohort patient bays or wards to isolate high patient-exhaled airborne viral loads. Regular maintenance and replacement of air cleaner filters are crucial for their effectiveness.
4. To stress the significance of avoiding overcrowding in various settings. Examples include students sitting at alternate desks in school classrooms, customers occupying every other Table in restaurants, or maintaining empty seats between passengers in public transport, cinemas, etc. This measure helps maintain physical distance and reduces the potential for close-contact transmission.

Overall, it fulfills its objectives by offering a comprehensive analysis of indoor airborne contamination, providing insights into effective mitigation strategies, and promoting the importance of maintaining healthy indoor environments for occupant well-being.

References

Adam, M. G., Tran, P. T. M., & Balasubramanian, R. (2021). Air Quality Changes in Cities During The COVID-19 Lockdown: A Critical Review. *Atmospheric Research*, 264: 105823. https://doi.org/10.1016/j.atmosres.2021.105823.

Aganovic, A., & Cao, G. (2019). Evaluation of Airborne Contaminant Exposure In A Single-Bed Isolation Ward Equipped with A Protected Occupied Zone Ventilation System. *Indoor and Built Environment*, 28(8): 1092–1103. https://doi.org/10.1177/1420326X18823048.

Aganovic, A., Steffensen, M., & Cao, G. (2019). CFD Study of the Air Distribution And Occupant Draught Sensation in A Patient Ward Equipped With Protected Zone Ventilation. *Building and Environment*, 162: 106279. https://doi.org/10.1016/j.buildenv.2019.106279.

Agarwal, N., Meena, C. S., Raj, B. P. et al., (2021). Indoor Air Quality Improvement in COVID-19 Pandemic: Review. *Sustainable Cities and Society*, 70: 102942. https://doi.org/10.1016/j.scs.2021.102942.

Al Assaad, D., Ghali, K., & Ghaddar, N. (2018). Effectiveness of Intermittent Personalized Ventilation Assisting A Chilled Ceiling For Enhanced Thermal Comfort and Acceptable Indoor Air Quality. *Building and Environment*, 144: 9-22. https://doi.org/10.1016/j.buildenv.2018.08.005.

Al Assaad, D., Ghali, K., & Ghaddar, N. (2019). Effect of Flow Disturbance Induced By Walking On The Performance of Personalized Ventilation Coupled with Mixing Ventilation. *Building and Environment*, 160: 106217. https://doi.org/10.1016/j.buildenv.2019.106217.

Al horr, Y., Arif, M., Katafygiotou, M., Mazroei, A., Kaushik, A., & Elsarrag, E. (2016). Impact Of Indoor Environmental Quality On Occupant Well-Being And Comfort: A Review of The Literature. *International Journal of Sustainable Built Environment*, 5, (1): 1–11. https://doi.org/10.1016/j.ijsbe.2016.03.006.

Ali Najeeb Alashaab, A., & Alamery, M. S. (2018). Investigation and Improvement the Thermal Comfort of the Air Conditioning Mosque at Hot - Dry Climate in Baghdad. *IOP Conference Series: Materials Science and Engineering*, 454(1). https://doi.org/10.1088/1757-899X/454/1/012154.

ALJABAIR, S., ALESBE, I., & ALKHALAF, A. (2023). CFD Modeling Of Influenza Virus Diffusion During Coughing And Breathing in A Ventilated Room. *Journal of Thermal Engineering*, 9(1): 127–137. https://doi.org/10.18186/thermal.1243491.

Al-Kayiem, H. H., Koh, K., Riyadi, T. W. B., & Effendy, M. (2020). A Comparative Review On Greenery Ecosystems And Their Impacts On Sustainability of Building Environment. *Sustainability*, 12(20): 1–26. https://doi.org/10.3390/su12208529.

Al-Kayiem, H. H., Mohammed, M. N., Kelly, K., Riyadi, T. W. B., & Effendy, M. (2023). Experimental Assessment and Development of Thermal Comfort Model for

Implication in Tropical Climate. *International Journal of Computational Methods and Experimental Measurements*, 11(1): 35–43. https://doi.org/10.18280/ijcmem.110105.

Al-Kayiem, H. H., Effendy, M., Riyadi, T. W. B., Kurnia, J. C., Morganadus, A. (2023). Influence of Life Green Wall and Roof Shedding on the Internal Thermal Condition of Buildings: Case Study in Malaysia. In: Ahmad, F., Al-Kayiem, H. H., King Soon, W. P. (eds) *ICPER 2020. Lecture Notes in Mechanical Engineering*. Springer, Singapore. https://doi.org/10.1007/978-981-19-1939-8_3.

Alkhalaf, M., Ilinca, A., & Hayyani, M. Y. (2023). CFD Investigation of Ventilation Strategies to Remove Contaminants from a Hospital Room. *Designs*, 7(1). https://doi.org/10.3390/designs7010005.

Berlanga, F. A., de Adana, M. R., Olmedo, I., Villafruela, J. M., San José, J. F. S., and Castro, F. (2018). Experimental Evaluation of Thermal Comfort, Ventilation Performance Indices and Exposure to Airborne Contaminant In An Airborne Infection Isolation Room Equipped With A Displacement Air Distribution System. *Energy and Buildings*, 158: 209–221. https://doi.org/10.1016/j.enbuild.2017.09.100.

Bluyssen, P. M., Fernandes, E. D. O., Groes, L., Clausen, G., Fanger, P. O., Valbjørn, O., Bernhard, C. A., and Roulet, C. A. (1996). European Indoor Air Quality Audit Project in 56 Office Buildings. *Indoor Air*, 6(4): 221-238. https://doi.org/10.1111/j.1600-0668.1996.00002.x.

Cao, G., Awbi, H. and Yao, R. et al., (2014). A Review of The Performance of Different Ventilation And Airflow Distribution Systems In Buildings. *Building and Environment*, 73: 171–186. https://doi.org/10.1016/j.buildenv.2013.12.009.

Cao, G., Nielsen, P. V., Jensen, R. L., Heiselberg, P., Liu, L., and Heikkinen, J. (2015). Protected Zone Ventilation and Reduced Personal Exposure to Airborne Cross-Infection. *Indoor Air*, 25(3): 307–319. https://doi.org/10.1111/ina.12142.

Cao, G., Storås, M. C. A., Aganovic, A., Stenstad, L-I., and Skogås, J.G. (2018). Do Surgeons and Surgical Facilities Disturb the Clean Air Distribution Close to A Surgical Patient in an Orthopedic Operating Room with Laminar Airflow? *American Journal of Infection Control*, 46(10): 1115-1122. https://doi.org/10.1016/j.ajic.2018.03.019.

Cheng, Z., Aganovic, A., Cao, G., & Bu, Z. (2021). Experimental and Simulated Evaluations Of Airborne Contaminant Exposure in a Room with a Modified Localized Laminar Airflow System. *Environmental Science and Pollution Research*, 28(24): 30642-30663. https://doi.org/10.1007/s11356-021-12685-4.

Cheong, K. W. D., & Phua, S. Y. (2006). Development of Ventilation Design Strategy for Effective Removal of Pollutant in the Isolation Room of a Hospital. *Building and Environment*, 41(9): 1161-1170. https://doi.org/10.1016/j.buildenv.2005.05.007.

Chow, T. T., Kwan, A., Lin, Z., & Bai, W. (2006a). A Computer Evaluation of Ventilation Performance in A Negative-Pressure Operating Theater. *Anesthesia and Analgesia*, 103(4): 913–918. https://doi.org/10.1213/01.ane.0000237404.60614.24.

Chow, T. T., Kwan, A., Lin, Z., & Bai, W. (2006b). Conversion of Operating Theatre From Positive to Negative Pressure Environment. *Journal of Hospital Infection*, 64(4): 371–378. https://doi.org/10.1016/j.jhin.2006.07.020.

Chuang, H. C., Lin, L-Y., Hsu, Y-W., Ma, C-M., Chuang, K-J. (2013). In-car Particles And Cardiovascular Health: An Air Conditioning-Based Intervention Study. *Science of the*

Total Environment, 452-453: 309–313. https://doi.org/10.1016/j.scitotenv.2013.02.097.

Commission on Environment, W. (1987). *Report of the World Commission on Environment and Development: Our Common Future Towards Sustainable Development 2*. Part II. Common Challenges Population and Human Resources.

De Giuli, V., Da Pos, O., & De Carli, M. (2012). Indoor Environmental Quality and Pupil Perception in Italian Primary Schools. *Building and Environment*, 56: 335-345. https://doi.org/10.1016/j.buildenv.2012.03.024.

Ding, E., Zhang, D., & Bluyssen, P. M. (2022). Ventilation Regimes of School Classrooms Against Airborne Transmission of Infectious Respiratory Droplets: A Review. *Building and Environment*, 207: 108484. https://doi.org/10.1016/j.buildenv.2021.108484.

Elsaid, A.M., & Ahmed, M.S. (2021). Indoor Air Quality Strategies for Air-Conditioning and Ventilation Systems with the Spread of the Global Coronavirus (COVID-19) Epidemic: Improvements and Recommendations. *Environmental Research*, 199: 111314. https://doi.org/10.1016/j.envres.2021.111314.

Fadaei, A. (2021). Ventilation Systems and COVID-19 Spread: Evidence from a Systematic Review Study. *European Journal of Sustainable Development Research*, 5(2): em0157. https://doi.org/10.21601/ejosdr/10845.

Ferrari, S., Blázquez, T., Cardelli, R., Puglisi, G., Suárez, R., and Mazzarella, L. (2022). Ventilation Strategies To Reduce Airborne Transmission of Viruses in Classrooms: A Systematic Review of Scientific Literature. *Building and Environment*, 222: 109366. https://doi.org/10.1016/j.buildenv.2022.109366.

Gilkeson, C. A., Camargo-Valero, M. A., Pickin, L. E., & Noakes, C. J. (2013). Measurement of Ventilation and Airborne Infection Risk in Large Naturally Ventilated Hospital Wards. *Building and Environment*, 65: 35–48. https://doi.org/10.1016/j.buildenv.2013.03.006.

Gola, M., Settimo, G., & Capolongo, S. (2019). Indoor Air Quality in Inpatient Environments: A Systematic Review on Factors that Influence Chemical Pollution in Inpatient Wards. *Journal of Healthcare Engineering*, Article ID 8358306. https://doi.org/10.1155/2019/8358306.

Gries, D., Schneider, F. B. (1995). Avoiding the Undefined by Under specification. Chapter in In: van Leeuwen, J. (eds) Computer Science Today. *Lecture Notes in Computer Science,* 1000. Springer, Berlin, Heidelberg. https://doi.org/10.1007/BFb0015254.

Health Building Note 00-01: General Design Guidance for Healthcare Buildings. (2014). www.gov.uk/government/collections/.

Karottki, D. G., Spilak, M., Frederiksen, M., Gunnarsen, L., Brauner, E. V., Kolarik B, Andersen ZJ, Sigsgaard T, Barregard L, Strandberg B, Sallsten G, Møller P, and Loft S. (2013). An Indoor Air Filtration Study In Homes Of Elderly: Cardiovascular And Respiratory Effects of Exposure to Particulate Matter. *Environ Health*, 12: 116. https://doi.org/10.1186/1476-069X-12-116.

Laouadi, A. (2022). A New General Formulation for the PMV Thermal Comfort Index. *Buildings*, 12(10): 1572. https://doi.org/10.3390/buildings12101572.

Li, T., Liu, Y., Wang, D., Shang, K., and Liu, J. (2015). Optimization Analysis on Storage Tank Volume in Solar Heating System. *Procedia Engineering*, 121: 1356-1364. https://doi.org/10.1016/j.proeng.2015.09.019.

Cao, G., Nielsen, P. V., Jensen, R. L., Heiselberg, P., Liu, L., and Heikkinen, J. (2015). Protected Zone Ventilation and Reduced Personal Exposure to Airborne Cross-Infection. *Indoor Air*, 25(30: 307-319. https://doi.org/10.1111/ina.12142.

Liu, Zhijian Liu, Wang, Y., Xia, X. et al., (2022). A Novel Design Strategy to Interrupt the Diffusion of Exhaled Aerosols: An Integrated Experimental-Numerical Investigation. *SSRN Electronic Journal*. 10.2139: ssrn.4166072. http://dx.doi.org/10.2139/ssrn.4166072.

Lu, Y., Oladokun, M., & Lin, Z. (2020). Reducing the Exposure Risk in Hospital Wards by Applying Stratum Ventilation System. *Building and Environment*, 183. https://doi.org/10.1016/j.buildenv.2020.107204.

Morawska, L. (2006). Droplet Fate in Indoor Environments, or Can We Prevent the Spread Of Infection? *Indoor Air*, 16(5): 335–347. https://doi.org/10.1111/j.1600-0668.2006.00432.x.

Mousavi, E.S., & Grosskopf, K.R. (2014). Ventilation Rates and Airflow Pathways in Patient Rooms: A Case Study of Bioaerosol Containment and Removal. *Annals of Occupational Hygiene*, 59(9): 1190–1199. https://doi.org/10.1093/annhyg/mev048.

Noorimotlagh, Z., Jaafarzadeh, N., Martínez, S. S., and Mirzaee, S. A. (2021). A Systematic Review of Possible Airborne Transmission of The COVID-19 Virus (SARS-Cov-2) In The Indoor Air Environment. *Environmental Research*, 193: 110612. https://doi.org/10.1016/j.envres.2020.110612.

Offermann, F. J., Eagan, A., Offermann, A. C. et al., (2016). Potential Airborne Pathogen Transmission in a Hospital with and Without Surge Control Ventilation System Modifications. *Building and Environment,* 106: 175–180. https://doi.org/10.1016/j.buildenv.2016.06.029.

Piscitelli, P., Miani, A., Setti, L. et al., (2022). The Role Of Outdoor And Indoor Air Quality In The Spread Of SARS-Cov-2: Overview and Recommendations by the Research Group on COVID-19 and Particulate Matter (RESCOP Commission). *Environmental Research*, 211: 113038. https://doi.org/10.1016/j.envres.2022.113038.

Sarbu, I., & Sebarchievici, C. (2013). Aspects of Indoor Environmental Quality Assessment in Buildings. *Energy and Buildings*, 60: 410–419. https://doi.org/10.1016/j.enbuild.2013.02.005.

Tang, J. W., Bahnfleth, W. P, Bluyssen, P. M. et al., (2021). Dismantling Myths on the Airborne Transmission of Severe Acute Respiratory Syndrome Coronavirus-2 (SARS-CoV-2). *Journal of Hospital Infection*, 110: 89–96. https://doi.org/10.1016/j.jhin.2020.12.022.

Thermal Environmental Conditions for Human Occupancy. (2004). Retrieved from www.ashrae.org.

Wang, Y., Zhai, C., Zhao, T., & Cao, Z. (2020). Numerical Study on Pollutant Removal Performance of Vortex Ventilation with Different Pollution Source Locations. *Building Simulation*, 13(6): 1373–1383. https://doi.org/10.1007/s12273-020-0632-3.

Yuan, F., Zhang, Y., Han, O., Li, A. et al., (2022). Thermal Comfort in Hospital Buildings – A Literature Review. *Journal of Building Engineering*, 45: 103463. https://doi.org/10.1016/j.jobe.2021.103463.

Zhang, Y., Han, O., Li, A., Hou, A., Olofsson, T., Zhang, L., and Lei, W. (2022). Adaptive Wall-Based Attachment Ventilation: A Comparative Study on Its Effectiveness in Airborne Infection Isolation Rooms with Negative Pressure. *Engineering*, 8: 130–137. https://doi.org/10.1016/j.eng.2020.10.020.

Chapter 6

Cranioplasty Materials for Human Skull Rehabilitation

Nareen Hafidh Obaeed[1,*]
and Wisam K. Hamdan[2]
[1]Production Engineering and Metallurgy Department,
University of Technology- Iraq, Baghdad, Iraq
[2]Biomedical Engineering Department University of Technology- Iraq, Baghdad, Iraq

Abstract

Accidental traumatic loads that fracture the skull or result in other injuries requiring a craniotomy, including the removal of a piece of the skull, may be applied to the human head. As a result, the removed part is replaced with an autologous bone or biocompatible material. Cranioplasty is one of the essential aims of sustainable development, which assures that everyone has access to health, justice, and prosperity. No one must be left behind, where the objective of cranioplasty is to protect the brain after cranial surgery. Furthermore, it will alleviate psychological difficulties while improving social performance. The search for acceptable skull bone substitutes has been a challenge for scientists worldwide. This review seeks to provide meta-data to present an aggregated picture of prospective trends for implant materials. It will serve as a fundamental reference point for future research using developing biomedical materials in cranial repair. The purpose of this review was to evaluate the role of cranial implants made of metallic, ceramic, and polymeric materials. The most used biocompatible materials, titanium alloy, polyetheretherketone, acrylic polymethylmethacrylate, and hydroxyapatite, were examined as part of

* Corresponding Author's Email: nareen.h.obaeed@uotechnology.edu.iq.

In: Advancements in Sustainability Systems
Editors: Hussain H. Al-Kayiem, Hasanain A. Abdul Wahhab and Raed A, Jessam
ISBN: 979-8-89113-571-0

the most significant research done in this area over the past five years. The present overview indicates several crucial areas for current and future research to improve and describe the crucial aspects of cranial implants and provide a starting point for understanding the capabilities and constraints of these common materials.

Keywords: cranial implant, cranioplasty, hydroxyapatite, PEEK, PMMA, Ti6Al4V

Abbreviations

PEEK	Polyetheretherketone.
PMMA	Polymethylmethacrylate.
DICOM	Digital Imaging and Communications in Medicine.
CT	Computerized Tomography.
HU	Hounsfield unit.
STL	Standard Tessellation Language.
EBM	Electron Beam Melting.
SPF	Superplastic Forming.
SPIF	Single Point Incremental Forming.
FEA	Finite Element Analysis.
ANN	Artificial neural network.
FFF	Fused filament fabrication.
PSIs	Patient-specific implants.
PLA	Polylactic acid.
CAD	computer-aided design.
TBI	traumatic brain injury.
HA	Hydroxyapatite.

1. Introduction

The rehabilitation of the skull bones is an enormous challenge due to the specificity of the topography of the skull from one person to another. Cranial implants are designed based on Digital Imaging and Communications in Medicine (DICOM) (Kim et al., 2023). Cranioplasty is a reconstructive surgical procedure that recreates the physiological functions of the cerebral skull and restores the structural integrity of craniofacial defects. The goal of cranioplasty is not only to eliminate the defect of the skull but also to improve the aesthetics, quality of life, and psychological state of the patient (Sharma et al., 2021; Alkhaibary et al., 2020).

Traumatic brain injuries are the leading cause of mortality and disability, affecting around 69 million people annually and placing a significant financial load on medical services (Jindal et al., 2021). Skull deformities are among the most common deformities that occur as a result of accidents arising from external trauma, swelling, decompression craniectomy or ruptured aneurysm, and terrorist operations, as depicted in Figure 1.

In cranioplasty, skull defects are corrected with autologous bone or artificial implants. In contrast, the purpose of treatment is to protect the brain, improve appearance, and reduce neurological symptoms (Sharma et al., 2020). The gold standard for cranioplasty is the use of autologous bone; however, when autologous bone is not an option due to fracture, infection, etc., or when an autologous cranioplasty has failed in the past, a synthetic craniofacial implant is employed. Polymethylmethacrylate (PMMA), titanium-based alloys, hydroxyapatite, and polyetherketone (PEEK) are among the most popular synthetic materials that have been employed. Regarding mechanical characteristics, biocompatibility, and osseointegration potential, each of these materials has specific benefits and drawbacks (Lewin, 2020). This article reviews and highlights the most common materials now used in the field of cranioplasty as well as offering a historical overview of the fabrication of cranial implants. All these materials have unique advantages and disadvantages in terms of mechanical characteristics, biocompatibility, and osseointegration possibility.

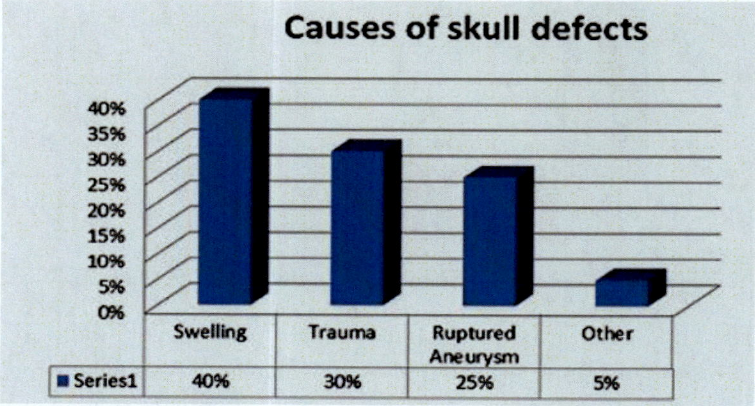

Figure 1. Causes of cranial defect.

2. Cranioplaty Procedure

It is a difficult task to design a suitable topographical replacement for a cranial defect rather than simply filling a hole (Cheng et al., 2018). There are several steps involved, and special programs that are used to repair the cranial defect. Reverse engineering is a key component of the proposed strategy that can be employed in the design of cranial implants, which involves obtaining patient data from a series of X-ray images around the skull collected at various angles combined to create a computerized tomography (CT) scan. To create an accurate 3D model, the quantity and thickness of slices are crucial factors (Dagang et al., 2021). To create a 3D structure, medical modeling software like Mimics (Materialise, Leuven, Belgium) could be used to import the Digital Imaging and Communications in Medicine (DICOM) data of the two-dimensional image slices from the CT scans, as shown in Figure 2.

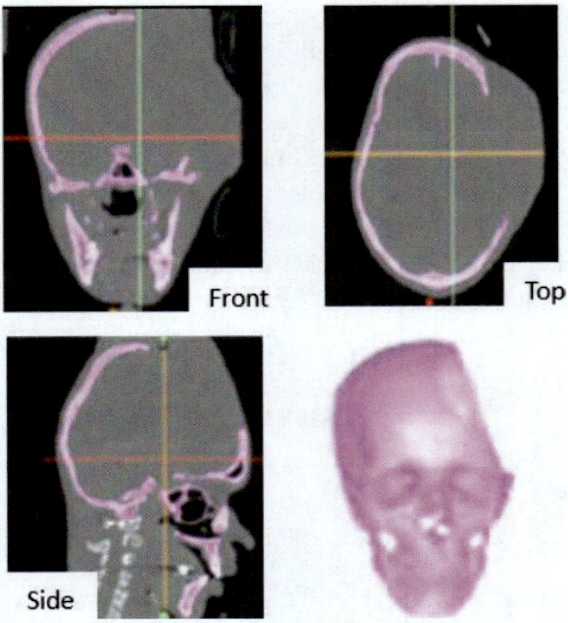

Figure 2. Three different orientations of CT scan images (Ameen et al., 2018).

The bone-specific is then marked using the Hounsfield unit (HU), and then a threshold selection is carried out. A Standard Tessellation Language (STL) file is then created for the implant's design after segmentation to contain the region of interest, as depicted in Figure 3. A healthy skull with marked experimental segmental damage serves as a model for implant design (Cheng et al., 2018; Schön et al., 2021).

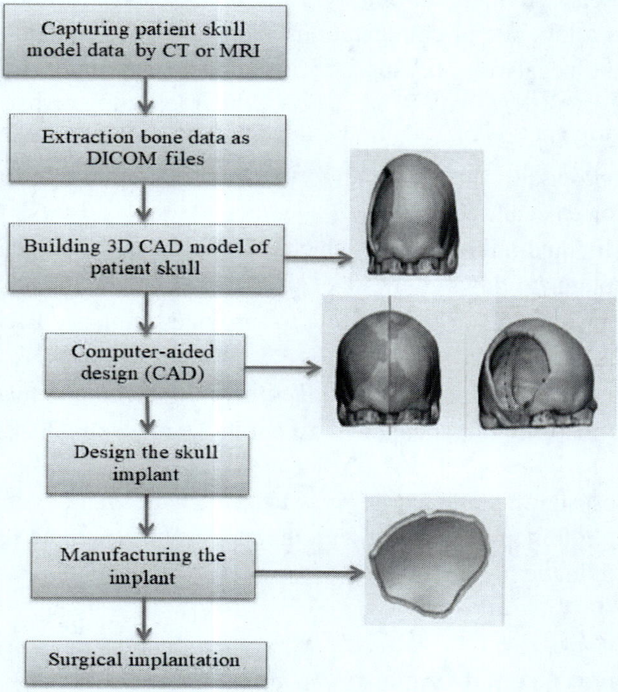

Figure 3. Production process for implantable patient specific.

3. Cranioplasty Materials

Although there are various materials available for cranial reconstruction, picking the right one for a specific defect is still challenging and ought to be performed with caution (Msallem et al., 2017).

The appropriate material for a cranioplasty must have characteristics that are similar to the bone it substitutes (Hatamleh, 2022; Aprianto et al., 2022). The following features should be evident in an appropriate implant:

- Non-toxic, non-geotropic, and having mechanical properties similar to the original bone (Petrovskaya et al., 2018; Bayirli et al., 2019). Metals like titanium, gold, silver and their alloys are among the earliest biomaterials utilized to repair defects in the skull. Even though many have been abandoned over time due to related medical issues, aluminum has been found to cause infections and epilepsy in patients having cranioplasty using this substance. Gold, in contrast, has a low rate of complications. This is costly, though, and silver is seen negatively because of its subpar mechanical qualities and elevated sensitivity to oxidation, which leads to corrosion (Jindal et al., 2021).

- Appropriate surface quality to provide implant biocompatibility (Correa et al., 2020).
- Individual micro profile, which offers a dependable means of fixing implants to flaws (Berretta et al., 2018).
- Appropriate surface quality to provide implant biocompatibility (Cheng et al., 2020).
- A cost-effective and quick manufacturing technique that guarantees the majority of patients will accept it (Cheng et al., 2020).

The most significant materials now used to create cranial implants will be highlighted in this review, along with the most significant and recent studies that focused on them.

3.1. Titanium Cranial Implants

Titanium alloy (Ti6Al4V) is one of the favorite biomaterials for many types of implants. It is suited for the application due to its high strength-to-weight ratio, biocompatibility, resistance to corrosion, high stiffness, nonmagnetic nature, and chemical stability (Martmez-Valencia et al., 2022; Bittredge et al., 2022). Titanium is used in the skull for attaching mechanisms like plates, screws, meshes, or hard plates. It is also utilized in conjunction with other materials like inert polymers or ceramic parts (Bittredge et al., 2022). Significant recent studies that repaired cranial defects with titanium alloy medical grade are reviewed and summarized below.

Ameen et al., 2018, outline a method for the design analysis and additive manufacturing-based manufacture of a unique, lightweight, rigid skull implant. An illustration of a case study involving the design, FEA analysis, and additive manufacturing employing electron beam melting (EBM) of a custom skull implant is shown. The implant material is a biocompatible, non-toxic titanium alloy called Ti6Al4V. The outcomes show that the manufacture of a thin, customized cranial implant for the Reconstruction of a skull defect employing EBM technology is successful. The constructed implant is suitable in terms of appearance and functionality and is strong and heavy enough to approach the weight of the removed bone.

Moiduddin et al., 2019, used Ti6Al4V powder to create the 3D model of the implant using electron beam melting (EBM). A tailored implant's biomechanical stability and strength have been assessed using a variety of tests. To retain both excellent aesthetics and anatomical accuracy, a comparative 3D analysis was conducted. The average yield strength and tensile strength of the made Ti6Al4V EBM sample, as well as the study of hardness, revealed results from the biomechanical investigation that also provided an acceptable result within the allowable range.

Palumbo et al., 2022, outline an effective design and fabrication process for a customized Ti6Al4V-ELI cranial prosthesis that has been demonstrated in vivo. For two unique and inventive sheet metal forming techniques, Superplastic Forming (SPF) and Single Point Incremental Forming (SPIF), used to make prostheses, a numerical/experimental design methodology has been suggested. Finally, the formative characteristics of the prostheses were assessed in terms of component geometry and thickness distribution.

Martmez-Valencia et al., 2022, adopted Finite Element Analysis (FEA) on a total of sixty different models. The material was put under intracranial pressure (ICP) conditions with 12 fixation points in the boundary conditions and a typical range (10 mm Hg). The designs were connected using an artificial neural network (ANN), which helped to achieve maximum displacement. The 5th to 95th percentiles, which represent the study's subject population, have a maximum displacement of 0.1 mm, and the best designs were produced using a generalized reduced gradient that minimizes the amount of material used.

Selvaraj et al., 2022, outline the approach utilized to build a specific cranium implant for a patient who suffered catastrophic injuries in a car accident. CAD design and 3D printing technology with selective laser melting (SLM) were used to create a 3D reconstruction of the cerebral vault. Recognizing that the prosthesis was in good condition, the duration of surgery was minimized, and any errors in the operating room were eliminated, and the implant was successfully inserted. The patient's aesthetics were restored, permitting the patient to securely undertake daily tasks with complete confidence.

Khan et al., 2023, reported the behavior of the planned implants is tested using titanium alloy (Ti6Al4V) in two scenarios: the impact of an external force by applying a load of 400N, which is like the force an adult man would strike with, and an increase in intracranial pressure. In addition to considering porosity as very important for osseointegration, the extent of porosity of these implants is compared. The study found that the octet, kelvin, and diamond lattice architectures produce good results and can be used to increase the implant's porosity while maintaining its load-bearing capacity. Van den Brink et al., 2023, described the use of a single implant made of 3D-printed Ti-6Al-4V grade 23 titanium alloy for skull restoration following oncological surgery using direct metal printing on ProX DMP320 printers. These patients were able to undergo reconstructive surgeries supported by implants without any issues.

3.2. Polyetheretherketone Cranial Implant

A team of English scientists created polyetheretherketone PEEK used for the first time in 1978. By the late 1990s, PEEK was replacing metal implant components, particularly in the orthopedic and trauma specialties. This polymer is a semi-crystalline thermoplastic with a chemical structure that offers mechanical strength and great thermal stability (Sarfraz et al., 2022). Due to its exceptional combination of high-temperature performance, chemical resistance, fatigue resistance, lightweight, high yield strength, stiffness, and durability, PEEK has subsequently been used in a variety of applications (Honigmann et al., 2018). The most significant recent studies that utilize PEEK as an implant in cranioplasty:

Sharma et al., 2020, characterized the geometric, morphological, and biomechanical features of hospital 3D-printed PEEK cranial implants. They report a quantification of 3D printed PEEK patient-specific implants PSI employing fused filament fabrication (FFF) sutures for skull reconstruction. The findings reveal that custom-made cranial implants are very dimensionally accurate and reproducible, and they also exhibit morphological resemblance that is clinically acceptable in terms of fit and continuity of characteristics. The studied implants failed at an average peak load (SD) of 798.38 ± 211.45 N, which was seen from a biomechanical perspective to have varied peak load values and discrete failure patterns.

Mian et al., 2022, provided a hybrid system that combines computed tomography with adaptive design (or simulation), computational analysis, and accuracy estimates. Anatomical reconstruction design methodology specific to each patient and defect is used to fabricate cranial implants made of PEEK in order to achieve the best results possible. Installation precision is acceptable, with a maximum outward deviation of 0.59 mm. This study reveals that labor-intensive changes are reduced while implant fit, stability, and strength are improved by using the 3D reconstruction technique and PEEK materials. Because it is quick and precise, the accepted design method can be used instead of other contemporary design methods.

Sharma et al., 2020, Examine PEEK PSIs made in-house for craniofacial Reconstruction and talk about the major difficulties encountered during the FFF printing process. An evaluation of the dimensional correctness of the manufactured implants was done after Taguchi's L9 orthogonal array was chosen for the non-thermal printing process parameters of layer thickness, infill rate, number of shells, and infill pattern. Deviations between (0.790 mm and 0.241 mm) were shown using the root mean square values. Different crystallinity zones were achieved, as well as horizontal lines. These findings imply that the dimensional accuracy of PSIs was within the clinically acceptable range. However, it is important to note that in order to produce implants with consistent crystallinity, the printing process needs to implement optimal thermal handling procedures.

Berretta et al., 2018, studied how the four production directions—horizontal, oblique, vertical, and horizontal inverted—affect the characteristics of PEEK laser-sintered skull implants for a specific patient. Researchers discuss dimensional accuracy in relation to the original design model, density change, and mechanical performance under uniaxial static pressure. Implants produced in the horizontal and inverted horizontal directions were found to have the maximum pressure resistance and the least departure from the design model. Regarding the inverted horizontal implant, the vertical design implant displayed lower design accuracy and a first failure rate of over 70%.

The fusion-filament technique (FFF) is utilized by Honigmann et al., 2018, to produce PEEK PSI. With the use of 3D printing, PEEK can build nearly any sophisticated design geometry that is not possible with conventional methods. Based on these preliminary findings, it may be concluded that PEEK can be successfully employed as a suitable biomaterial for the repair of surgical defects in a "biomimetic" design. This conclusion follows an investigation into the viability of PEEK production via 3D printing.

Chen et al., 2021, explored the temperature-dependent formability of PEEK in terms of producing the cranial plate utilizing incremental sheet forming ISF technique, a 3-axis desktop manufacturing system has been constructed. In the meantime, data on the forming force, temperature distribution, geometrical precision, and thermal qualities were gathered and examined. The results suggest that the ISF technique offers technological and financial advantages in cranial Reconstruction utilizing PEEK.

3.3. Polymethylmethacrylate Cranial Implant

Polymethylmethacrylate PMMA is a conventional transparent thermoplastic that was first utilized in cranioplasty in 1940 (Cheng et al., 2018). both during and after World War II, is one of the most employed allograft polymers that can be prefabricated or even molded (Sharma et al., 2020). However, some of the crucial studies to date that used PMMA as cranial implants are included in this chapter. The main contributed studies are outlined below.

Abd El-Ghani, 2018, explained the purposes, locations, and surgical techniques for polymethyl methacrylate (PMMA) implantation in seventy-five patients' individuals received these implants. The implant skull defect has covered any small to medium size (less than 8 cm). It has been made easier to reconstruct and avoid problems while applying for a concession. The results show that the Patients who used this implant experienced great cosmetic outcomes and no complications of the implant related. When done at the proper time and with the proper methods, implants produce cosmetic outcomes with little problems.

Chamo et al., 2020, performed an accuracy assessment of patient-specific implants (PSIs) manufactured from original molds. Ten PSIs were created from polymethylmethacrylate (PMMA) utilizing three-dimensional (3D) printing technology and computer-aided design (CAD). Fused deposition modeling was utilized to print a template from polylactic acid (PLA). These specialized implants' dimensions were measured, and they were compared to the original 3D-printed templates. This study demonstrates that high-fidelity PSI transcription can be achieved with an injection molding workflow. This workflow offers an advanced solution for skull reconstruction and continues to prove its worth in everyday clinical practice.

Kapri et al., 2020, Provided successful rehabilitation of a patient with a skull defect using an individual preoperative cranial prosthesis made of polymethyl methacrylate using a three-dimensional printed cranial prosthesis of PLA mold. This craniofacial prosthesis improved neurological symptoms through decompressive craniotomy and improved aesthetics and functionality.

Schön et al., 2021, demonstrated the manufacturing process for fabricating implants made of traditional PMMA using 3D printer-assisted molding technology for 16 recurring patients. The material of the mold was silicone heat resistant. Sixteen sequential patients undergo cranioplasty for a large skull defect with decompression surgery. The data were analyzed and evaluated according to preoperative, intraoperative, and postoperative.

Splavski et al., 2022, reported a 3D printed premade template manufactured by PolyJet additive technology was used to generate a unique PMMA implant to repair the calvarium in five consecutive patients who underwent craniectomy due to traumatic brain injury (TBI). All patients had a remarkable restoration of the symmetrical cranium's curve and contours of the skull. Clinically and cosmetically, all responses were favorable. Ashraf et al., 2022, printed frontotemporal parietal, bifrontal, frontal, and frontoparietal skull defects using polylactic acid PLA to consider implant size and adequacy before surgery. From the PLA implant, they designed a silicone mold that forms the PMMA implant. For one year, there were no postoperative complications, and the implants did not have to be removed. One of eight patients' implants required revision surgery because the orbital ridge had an unsatisfactory cosmetic result.

3.4. Hydroxyapatite Cranial Implant

Hydroxyapatite (HA) is a mixture of two calcium phosphates and water. Its composition and form resemble those of human bone. It has a hexagonal structure and promotes good osteointegration (Alkhaibary et al., 2020). The mechanical characteristics might vary greatly. It is because of structural variances brought on by the production process. The most current relevant studies on the use of hydroxyapatite in cranioplasty have been summarized in this chapter.

Spennato et al., 2020, used hydroxyapatite ceramic implants in eleven pediatric patients undergoing cranioplasty. An acceptable substitute for other cranioplasty procedures is hydroxyapatite ceramic implants. Implant mechanical resistance rises over time in areas where coaptation occurs well, and osteointegration is good.

Bruno et al., 2022, showed, by fibrosis and histological examinations, how the HA prosthesis can be incorporated after implantation. Five patients with primary HA craniotomies and second extractions a few years later are the subjects of the data collection. In all cases, isolated regions of osteolysis could be seen upon histological inspection of the newly generated lamellar/trabecular bone tissue fragmented with non-crystalline reticular tissue (HA prosthesis).

Amendola et al., 2022, aimed to investigate the efficacy of preoperative measures for reducing the likelihood of postoperative problems. Patients who needed thecal bone repair with HA prosthesis following post-traumatic decompressive surgery were eligible. Based on the results, they proposed in their study that the hydrophilic character of HA, as well as its structure that is a porous and rough surface, are significant factors in limiting early bacterial attachment and proliferation, hence lowering the likelihood of infection.

Martinez-Perez et al., 2022, reported a five-year retrospective review of sixty-nine patients who received translabyrinthine methods for lesion excision. Postoperative objective and subjective reconstruction markers, as well as postoperative complications, have been collected and examined. The employing of hydroxyapatite for cranium repair following translabyrinthine methods has increased patient satisfaction while decreasing cranial defects. Postoperative problems are consistent with the previous approaches described.

4. Selection of Peek, PMMA, and Titanium Alloy

Even though autologous bone is still considered the gold standard, there are occasions when it can't be utilized or is not available. The alternatives for cranial reconstruction have expanded as a result of improvements in alloplastic materials and cranioplasty procedures (Oliver et al., 2019). In terms of mechanical characteristics, biocompatibility, and osseointegration potential, each of these materials has specific benefits and drawbacks. One of the preferred biomaterials for many different types of implants is titanium alloy (Ti6Al4V). Due to its excellent strength-to-weight ratio, biocompatibility, corrosion resistance, high stiffness, nonmagnetic nature, and chemical stability, it is appropriate for the application (Moiduddin et al., 2019). The main issue with titanium implants is the stress shielding effect, which results in failure or bone loss because it has a higher elastic modulus than bone. In addition to high cost and limited malleability (Ameen et al., 2018).

As indicated in Table 1, titanium alloys have a significantly higher modulus of elasticity (110 GPa) than skull bones (14 GPa). Due to stress protection, bone resorption, and implant rupture, this significant difference between the two frequently causes implant malfunction (Agapovichev et al., 2018). When exposed to radiation, Grimal & Laugier, 2019 found that metal implants release dispersed rays that harm tissue. In addition to the substantial costs of manufacturing, researchers have been looking for new materials to replace the ones used in cranial implants to address the drawbacks and lessen biological issues that can arise after implant completion.

The use of PMMA, one of the most biocompatible alloplastic materials now on the market, for bone restoration dates back a long time. In addition to having the right mechanical qualities, it benefits from being inert, radiotransparent, nonmagnetic, easy to shape, and reasonably priced. Intraoperative molding of PMMA implants is also an option. PMMA provides advantages over intraoperative molding, including shorter surgery times, less blood loss and infection risk, and more satisfaction with the cosmetic outcome. The prefabrication procedure is simpler with PMMA than with other materials. The drawback of PMMA is that it might be harmful when employed in situ since it causes exothermic and toxic polymerization. In addition to potentially harming systemic organs, this polymerization could also leak monomers into the bloodstream and harm delicate peripheral structures (Sharma et al., 2020).

In cranial implant surgery, PEEK, a high-performance thermoplastic polymer, has grown in popularity. It is a fantastic material for cranial implants because it is biocompatible, efficient, and lightweight. Furthermore, PEEK implants' elasticity modulus is closer to the bone than that of metal implants, which can lessen stress shielding and enhance osseointegration. But compared to PMMA implants, shaping and contouring PEEK implants can be more challenging (Zhang et al., 2023).

The mineral composition of human bone is comparable to hydroxyapatite, and it is biocompatible, but it is fragile and prone to breaking. Therefore, it seems to be most suited for cranial defects that are small (van de Vijfeijken et al., 2018). Current suggestions argue against using hydroxyapatite as an inlay or on large defects (Hatamleh, 2022). PEEK and PMMA, two polymer materials utilized in cranioplasty, have superior resistance and compression features to hydroxyapatite (Alkhaibary et al., 2020). The limits of employing HA are its limited plasticity and expensive price.

Table 1. Characteristics of common cranioplasty materials

Material	Elastic Modulus (GPA)	Density (g/cm^3)	Thermal Conductivity (W/m·K)	Biocompatible
Titanium alloy	110	4.5	7.1	Yes
PMMA	3 - 5	1.18	0.167 - 0.25	Yes
PEEK	3 - 4	1.3	0.25 - 0.93	Yes
HA	3 - 6	3.16	1.5	Yes
Cortical Bone	14	1.6 – 2	0.68	

Regarding the issue of thermal conductivity, the implant should be manufactured from material comparable to bone because bone has a low thermal conductivity. This prevents the underlying brain tissue from being harmed by heat transfer from the atmosphere, especially within hot tropical regions.

Conclusion

The reconstruction of the skull's bones presents a difficult problem for the surgical team. Regarding the wide range of design techniques and materials available, the implant's design and choice of a suitable material are among the most significant variables in cranial reconstruction. Although titanium mesh is thought to have excellent biocompatibility, there are possible hazards associated with using it, particularly in regions with high temperatures like those in the Middle East and the sun belt. This is because titanium mesh is a heat-conductive material, which can cause scalp paresthesia. The thermoplastic materials that are shown are superior to titanium alloy because they have mechanical properties and strengths that are more like those of the skull bone and are less expensive to produce than expensive titanium implants.

References

Abd El-Ghani, W. M. A. (2018). Cranioplasty with Polymethyl Methacrylate Implant: Solutions of Pitfalls. Egyptian. *Journal of Neurosurgery*, 33(1). https://doi.org/10.1186/s41984-018-0002-y.

Agapovichev, A., Sotov, A., Kokareva, V., & Smelov, V. (2018). Possibilities and Limitations of Titanium Alloy Additive Manufacturing. *MATEC Web of Conferences*, 224. https://doi.org/10.1051/matecconf/201822401064.

Alkhaibary, A., Alharbi, A., Alnefaie, N., Almubarak, A. O., Aloraidi, A., and Sami Khairy, S. (2020). Cranioplasty: A Comprehensive Review of the History, Materials, Surgical Aspects, and Complications. *World Neurosurgery*, 139: 445–452. https://doi.org/10.1016/j.wneu.2020.04.211.

Ameen, W. Al-Ahmari, A., Mohammed, M.K., Abdulhameed, O., Umer, U., and Moiduddin, K. (2018). Design, Finite Element Analysis (FEA), and Fabrication of Custom Titanium Alloy Cranial Implant Using Electron Beam Melting Additive Manufacturing. *Advances in Production Engineering and Management*, 13(3): 267–278. https://doi.org/10.14743/apem2018.3.289.

Amendola, F. Vaienti, L., Carbonaro, R., Nataloni, A., Barbanera, A., Zingaretti, N., Pier, C. P., and Zanotti, B. (2022). Antibiotic Immersion of Custom-Made Porous

Hydroxyapatite Cranioplasty: A Multicentric Cohort Study. *Journal of Craniofacial Surgery,* 33(5): 1464–1468. https://doi.org/10.1097/SCS.0000000000008363.

Aprianto, D. R., Parenrengi, M. A., Utomo, B., Fauzi, A. Al, Subagio, E. A., & Suryawan, A. (2022). Comparison of Autograft and Implant Cranioplasty In *Pediatrics: A Meta-Analysis. Surgical Neurology International*, 13. https://doi.org/10.25259/SNI_1204_2021.

Ashraf, M., Choudhary, N., Kamboh, U. A., Raza, M. A., Sultan, K. A., Ghulam, N., Hussain, S. S., and Ashraf, N. (2022). Early Experience With Patient-Specific Low-Cost 3D-Printed Polymethylmethacrylate Cranioplasty Implants in a Lower-Middle-Income-Country: Technical Note and Economic Analysis. *Surgical Neurology International* 13. https://doi.org/10.25259/SNI_250_2022.

Bayirli, A., Ide, S., Orujalipour, I., Demir, O., Dursun, M., Ates, H. (2019). Nanoscale Structural Investigation on Ti-6Al-4V Implants Produced By Using Selective Laser Melting Technique: The Importance of Production Angle. *Journal of Materials Research and Technology*, 8(3): 2796-2801. https://doi.org/10.1016/j.jmrt.2019.04.019.

Berretta, S., Evans, K., & Ghita, O. (2018). Additive Manufacture of PEEK Cranial Implants: Manufacturing Considerations Versus Accuracy and Mechanical Performance. *Materials and Design*, 139: 141-152. https://doi.org/10.1016/j.matdes.2017.10.078.

Bittredge, O. Hassanin, H., El-Sayed, M. A., Eldessouky, H. M., Alsaleh, N. A., Alrasheedi, N. H., Essa, K. and Ahmadein, M. (2022). Fabrication and Optimisation of Ti-6Al-4V Lattice-Structured Total Shoulder Implants Using Laser Additive *Manufacturing. Materials*, 15(9). https://doi.org/10.3390/ma15093095.

Bruno, Z., Angelo, N., Riccardo, S., Nicola, Z., Stefano, P., Camillo, P. P., Federico, N., Carlotta, M. (2022). Case Report Custom-made Hydroxyapatite Cranioplasty: Radiological and Histological Evidence of Bone - Biomaterial Osteointegration in Five Patients. *Asian J Neurosurg* 15(1): 198-203. http://doi.org/10.4103/ajns.AJNS_208_19.

Chamo, D., Msallem, B., Sharma, N., Aghlmandi, S., Kunz, C., and Thieringer, F. M. (2020). Accuracy Assessment of Molded, Patient-Specific Polymethylmethacrylate Craniofacial Implants Compared to Their 3D Printed Originals. *Journal of Clinical Medicine,* 9(3). https://doi.org/10.3390/jcm9030832.

Chen, L.-F., Chen, F., Gatea, S., and Ou, H. (2021). PEEK Based Cranial Reconstruction Using Thermal Assisted Thermal Assisted Incremental Sheet Forming. In *Proceedings of the Institution of Mechanical Engineers, Part B: Journal of Engineering Manufacture*, 236(6,7). https://doi.org/10.1177/09544054211045904.

Cheng, C-H., Chuang, H-Y., Lin, H-L., Liu, C-L., and Yao, C-H. (2018). Surgical Results of Cranioplasty Using Three-Dimensional Printing Technology. *Clinical Neurology and Neurosurgery*, 168: 118-123. https://doi.org/10.1016/j.clineuro.2018.03.004.

Cheng, Z., Li, Y., Xu, C., Liu, Y., Ghafoor, S., Li, F. (2020). Incremental Sheet Forming Towards Biomedical Implants: A Review. *Journal of Materials Research and Technology*, 9(4): 7225-7251. https://doi.org/10.1016/j.jmrt.2020.04.096.

Correa, D. R. N., Rocha, L. A., Donato, T. A. G., Sousa, K. S. J., Grandini, C. R., Afonso, C. R. M., Doi, H., Tsutsumi, Y., Hanawa, T. (2020). On the Mechanical

Biocompatibility of Ti-15Zr-Based Alloys For Potential Use as Load-Bearing Implants. *Journal of Materials Research and Technology* 9(2): 1241–1250. https://doi.org/10.1016/j.jmrt.2019.11.051.

Dagang, W., Mahmud, J., Manan, N. F. A., & Abdullah, A. H. (2021). The Reconstruction Of Three-Dimensional (3D) Model of The Right Parietal-Temporal Implant. *AIP Conference Proceedings*, 2344. https://doi.org/10.1063/5.0047476.

Grimal, Q., & Laugier, P. (2019). Quantitative Ultrasound Assessment of Cortical Bone Properties Beyond Bone Mineral Density. *Irbm*, 40(1): 16-24. https://doi.org/10.1016/j.irbm.2018.10.006.

Hatamleh, M. M. (2022). Contemporary Review on Craniectomy and Cranioplasty; Part 2: Material Selection and Plate Manufacture. *Journal of Craniofacial Surgery* 33(3): 842–845. https://doi.org/10.1097/SCS.0000000000008040.

Honigmann, P., Sharma, N., Okolo, B., Popp, U., Msallem, B., and Thieringer, F. M. (2018). Patient-Specific Surgical Implants Made of 3D Printed PEEK: Material, Technology, and Scope of Surgical Application. *BioMedical Research International*, Article ID 4520636. https://doi.org/10.1155/2018/4520636.

Jindal, P., Bharadwaja, S., Rattra, S., Pareek, D., Gupta, V., Breedon, P., Reinwald, Y., and Juneja, M. (2022). Optimizing Cranial Implant and Fixture Design Using Different Materials in Cranioplasty. In *Proceedings of the Institution of Mechanical Engineers, Part L: Journal of Materials: Design and Applications*, 237(1). https://doi.org/10.1177/14644207221104875.

Jindal, S., Manzoor, F., Haslam, N., and Mancuso, E. (2021). 3D Printed Composite Materials For Craniofacial Implants: Current Concepts, Challenges and Future Directions. *International Journal of Advanced Manufacturing Technology*, 112(3-4): 635-653. https://doi.org/10.1007/s00170-020-06397-1.

Kapri, A., Kumari, P., and Sethna, G. (2020). Rehabilitation of a Cranial Defect with A Preoperatively Customized Polymethyl-Methacrylate Prosthesis Using 3-Dimensional Printed Polylactic Acid Mold: A Case Report. *IP Annals of Prosthodontics and Restorative Dentistry,* 6(2): 105–109. https://doi.org/10.18231/j.aprd.2020.022.

Khan, M. Z., Bhaskar, J., and Kumar. A. (2023). Design and Analysis of Strut-Based Lattice Structure Cranial Implant. *Journal of Mechanical Engineering and Sciences*, 17(1): 9307–9314. https://doi.org/10.15282/jmes.17.1.2023.1.0735.

Kim, C. N. T., Binh, C. X., Dung, V. T., and Toan, T. V. (2023). Design and Mechanical Evaluation of a Large Cranial Implant and Fixation Parts. *Interdisciplinary Neurosurgery: Advanced Techniques and Case Management*, 31. https://doi.org/10.1016/j.inat.2022.101676.

Lewin, S. (2020). *Functional Aspects of Cranial Implants Mechanical and Regenerative Properties*. Ph. D. Thesis. Uppsala University, Sweden. http://urn.kb.se/resolve?urn=urn:nbn:se:uu:diva-409756.

Martinez-Perez, R., Kunigelis, K. E., Ward, R. C., Ung, T. H., Arnone, G. D., Cass, S. P., Gubbels, S. P., and Youssef, A. S. (2022). Hydroxyapati Hydroxyapatitete Cement Cranioplasty for Reconstruction of Translabyrinthine Approach: Aesthetic Results, Long-Term Satisfaction, Quality of Life, and Complications. *Acta Neurochirurgica,* 164(3): 669-677. https://doi.org/10.1007/s00701-021-05024-6.

Martmez-Valencia, M. I., Navarro, C., Vázquez-López, J., Hernández-Arellano, J., Jiménez-García, J. and Díaz-León, J. (2022). Optimization of Titanium Cranial Implant Designs Using Generalized Reduced Gradient Method, Analysis of Finite Elements, and Artificial Neural Networks. *Revista Internacional de Metodos Numericos Para Calculo y Diseno En Ingenieria*, 38(3). https://doi.org/10.23967/j.rimni.2022.06.004.

Mian, S. H., Moiduddin, K., Elseufy, S. M., and Alkhalefah, H. (2022). Adaptive Mechanism for Designing a Personalized Cranial Implant and Its 3D Printing Using PEEK. *Polymers*, 14(6). https://doi.org/10.3390/polym14061266.

Moiduddin, K., Mian, S. H., Umer, U., and Alkhalefah, H. (2019). Fabrication and Analysis of A Ti6Al4V Implant for Cranial Restoration. *Applied Sciences* 9(12). https://doi.org/10.3390/app9122513.

Msallem, B., Beiglboeck, F., Honigmann, P., Claude, J., and Florian, T. (2017). Craniofacial Reconstruction by a Cost-Efficient Template-Based Process Using 3D printing. *Plastic and Reconstructive Surgery - Global Open*, 5(11). https://doi.org/10.1097/GOX.0000000000001582.

Oliver, J. D., Joseph, B., Amjed, A. G., Vyas, K. S., and Basel, S. (2019). Alloplastic Cranioplasty Reconstruction: A Systematic Review Comparing Outcomes with Titanium Mesh, Polymethyl Methacrylate, Polyether Ether Ketone, and Norian Implants in 3591 Adult Patients. *Annals of Plastic Surgery*, 82(5S Suppl 4), S289-S294. https://doi.org/10.1097/SAP.0000000000001801.

Palumbo, G., Ambrogio, G., Crovace, A., Piccininni, A., Cusanno, A., Guglielmi, P., De Napoli, L., and Serratore, G. (2022). A Structured Approach for the Design and Manufacturing of Titanium Cranial Prostheses via Sheet Metal Forming. *Materials*, 12(2). https://doi.org/10.3390/met12020293.

Petrovskaya, T. S., Toropkov, N. E., Mironov, E. G., & Azarmi, F. (2018). 3D Printed Biocompatible Polylactide-Hydroxyapatite Based Material for Bone Implants. *Materials and Manufacturing Processes*, 33(16): 1899-1904. https://doi.org/10.1080/10426914.2018.1476764.

Sarfraz, S., Mäntynen, P-H., Laurila, M., Rossi, S., Leikola, J., Kaakinen, M., Suojanen, J. and Reunanen, J. (2022). Comparison of Titanium and PEEK Medical Plastic Implant Materials for Their Bacterial Biofilm Formation Properties. *Polymers*, 14(18). https://doi.org/10.3390/polym14183862.

Schön, S. N., Skalicky, N., Sharma, N., Zumofen, D. W., and Thieringer, F. M. (2021). 3D-Printer-Assisted Patient-Specific Polymethyl Methacrylate Cranioplasty: A Case Series of 16 Consecutive Patients. *World Neurosurgery*, 148: e356–e362. https://doi.org/10.1016/j.wneu.2020.12.138.

Selvaraj, S., Dorairaj, J., and M, S. (2022). 3D Cranial Reconstruction Using Titanium Implant - A Case Report. *African Health Sciences*, 22(3): 383-390. https://doi.org/10.4314/ahs.v22i3.41.

Sharma, N., Aghlmandi, S., Cao, S., Christoph Kunz, C., Honigmann, P., and Thieringer, F. M. (2020). Quality Characteristics and Clinical Relevance of In-House 3D-Printed Customized Polyetheretherketone (PEEK) Implants For Craniofacial Reconstruction. *Journal of Clinical Medicine*, 9(9): 1-17. https://doi.org/10.3390/jcm9092818.

Sharma, N., Aghlmandi, S., Dalcanale, F., Seiler, D., Zeilhofer, H-F., Honigmann, P., and Thieringer, F. M. (2021). Quantitative Assessment of Point-of-Care 3D-Printed Patient-Specific Polyetheretherketone (PEEK) Cranial Implants. *International Journal of Molecular Sciences*, 22(16). https://doi.org/10.3390/ijms22168521.

Spennato, P., Canella, V., Aliberti, F., Russo, C., Ruggiero, C., Nataloni, A., Lombardo, M. and Cinalli, G. (2020). Hydroxyapatite Ceramic Implants For Cranioplasty in Children: A Retrospective Evaluation of Clinical Outcome and Osteointegration. *Child's Nervous System,* 36(3), 551-558. https://doi.org/10.1007/s00381-019-04423-6.

Splavski, B., Lakicevic, G., Kovacevic, M., & Godec, D. (2022). Customized Alloplastic Cranioplasty of Large Bone Defects by 3D-Printed Prefabricated Mold Template After Post-traumatic Decompressive Craniectomy: A Technical Note. *Surgical Neurology International,* 13. https://doi.org/10.25259/SNI_1239_2021.

van de Vijfeijken, S. E. C. M., Münker, T. J. A. G., Spijker, R., Karssemakers, L. H. E., Vandertop, W. P., Becking, A. G., and Ubbink, D. T. (2018). Autologous Bone Is Inferior to Alloplastic Cranioplasties: Safety of Autograft and Allograft Materials for Cranioplasties, a Systematic Review. *World Neurosurgery*, 117: 443-452.e8. https://doi.org/10.1016/j.wneu.2018.05.193.

van den Brink, E. J. C., Grinwis, G. C. M., Willemsen, K., Driessen, F., Boroffka, S. A. E. B., and Meij, B. P. (2023). Additive Manufacturing of Titanium Implants for Skull Reconstruction in 2 Dogs after Bone Tumour Excision. *VCOT Open*, 06(01): e61-e66. https://doi.org/10.1055/s-0042-1758679.

Zhang, J., Su, Y., Rao, X., Pang, H., Zhu, H., Li, L., Chen, L., Li, D., He, J., Peng, J., and Jiang, J. (2023). Additively Manufactured Polyether Ether Ketone (PEEK) Skull Implant as an Alternative to Titanium Mesh In Cranioplasty. *International Journal of Bioprinting*, 9(1): 173-180. https://doi.org/10.18063/ijb.v9i1.634.

Chapter 7

Radiation Passive Technology for Photovoltaic Cell Cooling

Saddam K. Shabeeb[1,*]
and Muna S. Kassim[1]
[1]Mechanical Engineering Department, College of Engineering,
Mustansiriyah University, Baghdad, Iraq

Abstract

The significant obstacle limiting the PV solar cell performance is the high cell temperature associated with the use of photovoltaic modules, a type of equipment that converts solar radiation into electrical energy. It has been discovered that as the temperature of the solar module rises by one degree Celsius, the electrical output power of crystalline silicon photovoltaic cells drops by 0.2 - 0.5%. The absorption of lost photons, which solar cells do not benefit from in conversion into electricity and whose energy makes up about 20% of the total solar radiation, is one of the most significant sources of heat for solar cells. If absorbed by solar cells, this energy eventually transforms into heat. The maximum power output of a solar cell drops as the temperature rises because temperature influences a photovoltaic cell's open circuit voltage; thus, it is important to keep the photovoltaic panel as cool as possible. To effectively lower the temperature of solar cells without using energy or bulky components, radiation cooling is a feasible option. Commercial solar cells, such as silicon cells, and the materials used to enclose solar cells (like glass) already have high heat emissivity, indicating that they can be self-cooled through radiative cooling. The solar panels are cooled by increasing heat emission or spectrum selectivity for the rays that fall on them. Since solar

---

* Corresponding author email: saddamaltamimi1@gmail.com.

In: Advancements in Sustainability Systems
Editors: Hussain H. Al-Kayiem, Hasanain A. Abdul Wahhab and Raed A, Jessam
ISBN: 979-8-89113-571-0

panels only use a small portion of the electromagnetic spectrum, between 0.3 and 1.1 μm, excess radiation outside that range heats up the panels. Spectrum selectivity works by reflecting the unwanted rays and letting the useful ones through. To learn more about the benefits and capabilities of this sort of cooling, we will explore literature that addresses the topics of radiative cooling and solar spectrum control technologies.

Keywords: photovoltaic, passive cooling, radiant cooling, spectrally selective

Abbreviations

PV	Photovoltaic
FF	Fill Factor
UV	Ultraviolet radiation
VI	Visible radiation
IR	Infrared radiation
CdTe	Cadmium telluride
AM1.5	Air Mass (Standard Spectrum)
STC	Standard test condition
Cp	specific heat
RC	radiative cooling

Nomenclature

A	Surface Area of solar cell
T	Temperature
H	Latent heat of fusion
Isc	Short Circuit Current
Voc	Open Voltage Current
h	Planck constant
V	Frequency of the photon
Cp	Specific heat
G	Solar irradiance

Greek Symbols

λ	Wavelength of incoming light
ρ	Density (solid/liquid)
τ	Transmittance of the glass cover
η	Efficiency.

1. Introduction

Compared to fossil fuel reserves, the sun continues to be the most significant source of energy available, even though only a small amount reaches the Earth's surface. Solar photovoltaic systems are among the renewable energy technologies evolving quickly and are expected to play a significant role in the future. Solar radiation measures the intensity of sunlight energy received at every point on Earth and is given in watts per square meter (W/m^2). Photovoltaic cells are semiconductors that turn sunlight into energy and rely on solar radiation.

The open circuit voltage (Voc) declines, and the short circuit current (Isc) rises as solar radiation levels rise (Gu et al., 2020), increasing the power the PV panel can produce. The photovoltaic cell's temperature significantly impacts the rate at which photons are produced (Lupu et al., 2018), which in turn causes the reverse saturation current to grow quickly and narrow the band gap, altering the electrical characteristics of the photovoltaic cell (Radziemska, 2003). Due to the negative temperature coefficient (-0.5%/°C of temperature increase), the output power of the PV panel falls as the temperature rises, indicating a decrease in efficiency (measured in terms of power output) (Raval et al., 2014). A cooling system is necessary to lower the temperature of the PV panel because it operates more effectively at lower temperatures (Homadi et al., 2020). Numerous investigations have been conducted using various solar panel cooling techniques (Koohestani et al., 2023). To highlight the key characteristics of this kind of cooling technique, we shall review the works that addressed radiative cooling.

The rate at which the Earth receives energy from the sun, 81016 W, is over 10,000 times greater than the amount the world now needs. Solar energy is, therefore, a vital and substantial renewable energy source (Krotkus et al., 2023). Solar energy conversion systems are different from other systems in that it is challenging to control the amount of readily available energy because it is constantly changing and influenced by several factors. The most important of which are (Mohamed, 2010).

- The space between the Earth and the sun varies due to the Earth's eccentric orbit around the sun.
- The tilt of the Earth's rotating axis: The Earth's axis is inclined 23.5^{o} from its orbit around the sun, which disperses solar energy on the Earth's surface, changing the duration of the day and night and the four seasons.
- Atmosphere clarity: The gases surrounding the planet, as well as dust particles and vapors associated with the atmosphere, scatter solar radiation as it flows through the atmosphere.
- Geographical location: Generally, places with latitudes near the equator get more solar radiation than others.
- The panel's location: The solar concentrator in an area densely populated with buildings and trees may prevent direct sunlight from reaching it.
- Panel orientation: screens that follow the sun get more solar energy than others, and screens positioned in the south receive more solar radiation than screens set horizontally in northern hemisphere locations.
- Season: Because of the extended duration of the day and the night's brevity, solar radiation is more plentiful in the summers compared to other seasons, particularly in the Middle East.
- Weather conditions: Clouds, which may obstruct more than 50% of total sunlight (direct and diffuse), are the most critical weather elements influencing solar radiation.

Table 1. The solar electromagnetic spectrum

Solar radiation spectrum	Percentage	wavelength
Visible light	43%	380-780 nm
Infrared (IR) spectrum	49%	780- 2600 nm
Ultraviolet (UV) radiation	7%	200–380 nm
γ rays	%1	less than 0.01 nm
X rays		0.01 – 10 nm
Radio waves		1×10^{6} - 1×10^{16} nm

The solar spectrum is separated into three distinct parts (Oslo, 2009). Solar radiation is emitted at different wavelengths across the electromagnetic spectrum. These radiative spectrums were explained in Table 1 with the percentage of each solar component's wavelengths.

2. Photovoltaic Cells Technology

The technology of converting solar energy into electrical energy provides a solution to the problem of unavailability of electrical energy, especially in remote places, as well as clean and available energy, so photovoltaic technology is one of the most important alternative sources.

2.1. Photovoltaic Principles

Photovoltaic technology has the following features.

- The energy source is free. Thus, it is sustainable and will never be depleted.
- Clean, quiet, and dependable.
- Solar power requires huge spaces as well as an unattractive view.

Sunlight comprises packets of energy known as photons, which carry varying amounts of energy proportional to the spectrum of wavelengths of light. The equation that expresses the energy is (Olita, 2012),

$$E = h\,V \tag{1}$$

where h stands for the Planck constant, V is the photon's frequency.

A light beam's photon energy diminishes according to the wavelength's height. Some materials, such as semiconductors, can absorb a limited energy range of photons when exposed to light. In contrast, other photons may pass through the material or be reflected without being absorbed. Various semiconductors have various optical absorption coefficients.

When a photon enters an illuminated substance, its energy is transferred to an electron. When this energy exceeds the electron binding energy, the electron is ejected from the ground energy state, forming an electron-hole pair as depicted in Figure 1. Suppose the photocell is connected to a load. In that case, the electric field produced by the p-n junction then separates the holes and electrons, resulting in a potential difference and current (electricity) flowing through an external circuit (Olita, 2012).

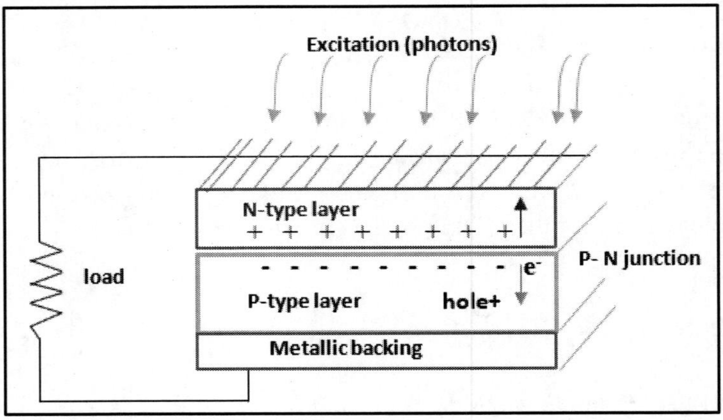

Figure 1. Photovoltaic P-N junction exposed to light.

2.2. The Optimal Wavelength Range for Photovoltaic Cells

The best wavelength range for photoelectric conversion in solar cells is between 0.3 and 1.1 μm (B. Zhao et al., 2020), as depicted in Figure 2.

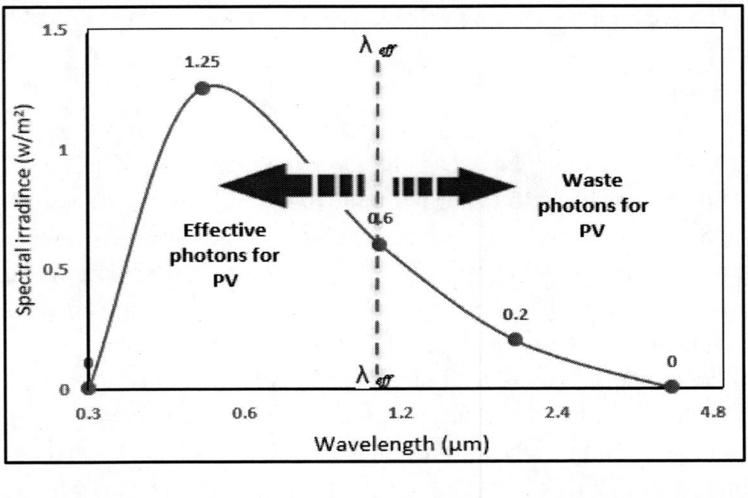

Figure 2. Schematic of effective wavelength for PV conversion with normalized AM 1.5 solar spectrum.

The parasitic absorption of wavelengths outside this range causes a rise in the temperature of the photovoltaic cell, which leads to low photovoltaic conversion efficiency. This is because the large drop in Voc more than compensates for the marginal rise in Isc (Dubey et al., 2013). Operating temperatures can sometimes reach values well above 325 K due to the absorption of photons outside the required range. A relative efficiency decreases of about 0.45% can be expected for a solar cell built of crystalline silicon for every 1.0 K increase in temperature (Perrakis et al., 2020).

The measurement most significantly impacted by an increase in cell or unit temperature is the open circuit voltage of a photovoltaic unit (PV) or solar cell. This is because the open circuit voltage is inversely proportional to the amount of light allowed to enter the cell, and the resistance in the circuit rises as a function of temperature because the speed of the electrons increases. The components used to make solar cells can exhibit different properties at various temperatures. As a result, the photovoltaic system must be constructed with a thorough grasp of the components that make up the photovoltaic panels and the general temperature pattern of its surroundings. This is because a PV system's performance can be impacted by various factors, including temperature (Agyekum et al., 2021). When attempting to characterize the temperature dependence of PV module performance, the temperature coefficient of a given material is used as the primary variable. The value of this coefficient is given as a decrease in voltage for each increase in the unit's temperature. For example, polycrystalline photovoltaic units witness a decrease in voltage of 0.4/°C (Kalaiselvan et al., 2018). Table 2 shows the temperature coefficient for different types of cells. Figure 3 also shows how the increase in temperature affects the amount of energy that can be extracted from a photovoltaic cell made of various materials.

Table 2. The performance of several different kinds of cells and their temperature coefficient

Type	STC performance (%)	Temperature Coefficient
Monocristalline Si (m-si)	12.5-15	-0.4 to -0.5
Polycristalline Si (p-si)	11-14	-0.4 to -0.5
Amorphous-s (a-Si)	11-13	-0.35 to -0.38
CiGS	10-13	-0.32 to -0.36
CdTe	9-12	-0.25

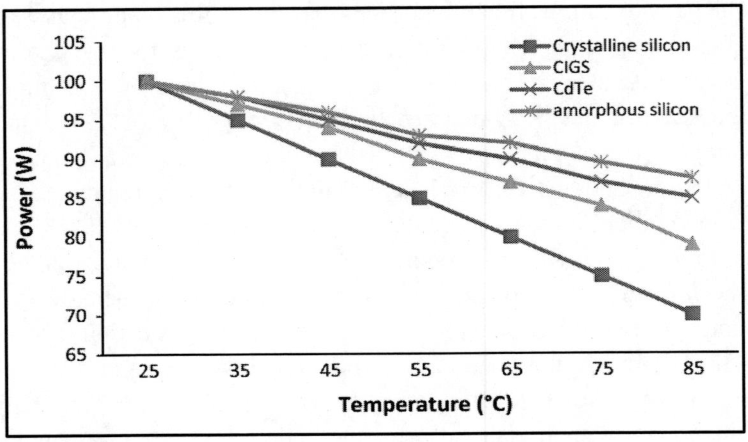

Figure 3. Temperature effect on different PV materials (Sun et al., 2017).

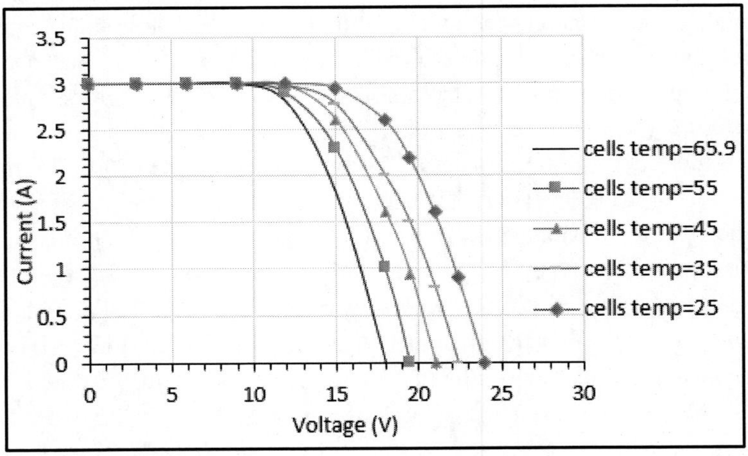

Figure 4. The I-V Curve of the PV panel at constant solar irradiance, 1000 W/m^{2}, and different temperatures of the PV panel.

Agroui, 2012 examined the performance of polycrystalline photovoltaic modules under changing temperatures and radiation levels. The experiments were conducted at temperatures ranging from 61 to 75 °C, while the radiation levels ranged from 780 to 1250 watts per square meter. It was observed that under these working settings, the efficiency of the units decreased by approximately 35%, and the maximum output power decreased by approximately 18% compared to the results obtained under normal test conditions.

Figure 4 shows the relationship between the voltage and current of solar cells when exposed to different temperatures, where the response of electric current flux is viewed as a function of changing the voltage of the photovoltaic panel when the solar irradiation is constant at 1000 W/m^2 and the temperature changes.

3. PV Cooling Techniques

Radiation that falls outside the range of wavelengths on which the solar cell operates leads to a rise in the surface temperature of the PV module, which reduces its efficiency. PV efficiency represents the ratio between the amount of electricity produced by the photovoltaic module and the amount of sunlight that hits it, and to maintain the electrical performance of the photovoltaic module at a level Well, it is important to use the correct cooling method to reduce the surface temperature of the PV, making it last longer (Mahdavi et al., 2022).

The solar cell cooling techniques are divided into active and passive cooling techniques, both of which consist of several methods, as shown in Figure 5.

3.1. Radiative Cooling

This part discusses a group of current scientific studies on cooling solar cells using radiative cooling methods. Outer space has great potential to be used as a heat sink because it has a very low temperature (about 3 degrees Kelvin). According to Stefan Boltzmann's law, all objects emit radiation at temperatures above absolute zero, and the higher the temperature of an object, the more radiation it emits (Sato & Yamada, 2019). Since the surface temperature of the Earth and most terrestrial bodies is about 300°K, the Earth in the infrared (MIR) wavelength range (4–25 μm) (Zhao et al., 2019) emits most of the thermal radiation.

Because of the large temperature difference, the Earth's heat is transmitted into space through thermal radiation, which ultimately maintains the Earth's temperature to remain habitable. Water vapor and atmospheric gases reduce this heat transfer. Still, in the wavelength range of 8–13 μm (that is, the atmospheric window), the atmosphere is very transparent and allows heat to pass through radiation, as shown in Figure 6. Exploring the atmospheric window to cool off solar cells by radiative cooling of wavelengths within the window range is possible. In addition, it is possible to exploit a window. Atmospheric cooling of solar cells by radiative cooling of wavelengths within the window range (Ahmed et al., 2021).

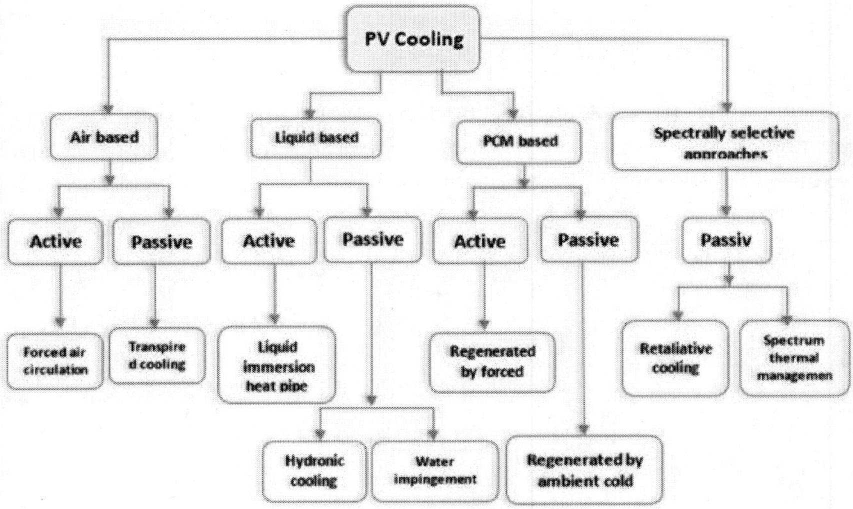

Figure 5. A summary of the different PV cooling methods.

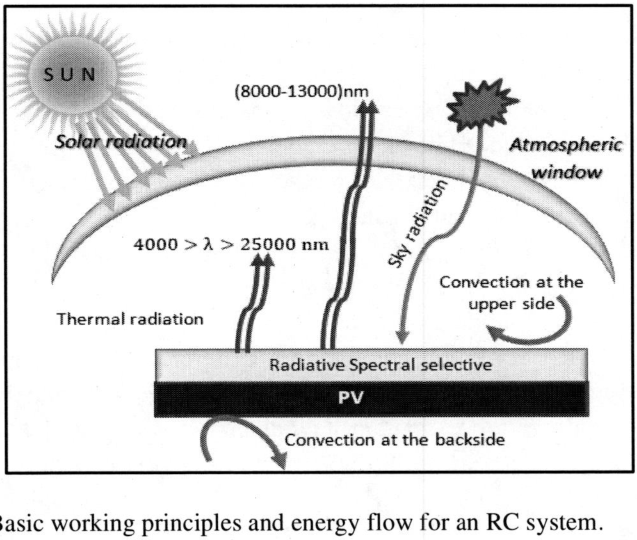

Figure 6. Basic working principles and energy flow for an RC system.

Daytime photovoltaic passive radiative cooling automatically cools the surface by reflecting sunlight and radiating heat back into the cool outer space. The main function of radiative cooling is to enhance solar cells' heat emission or reduce the absorption of waste photons by solar energy (Dwivedi et al., 2020). For cells that cannot create a pair of electron holes, this strategy relies on adding a transparent coolant on top of the solar cells to increase the thermal emission of the solar cells while maintaining the efficiency of solar energy absorption (Mandal et al., 2018). The transparent coolant should have a high transmittance in the wavelength range associated with solar radiation and a large heat emission in the wavelength range associated with central infrared radiation (Li et al., 2021). It is possible to cool solar panels by reflecting their heat to a cooler outer space through a transparent window for electromagnetic waves extending between 8 and 13 μm, corresponding to the peak wavelengths of heat radiation. Radiative cooling attracts efforts to improve energy efficiency because it is completely passive and does not require any power input (Rephaeli et al., 2013).

A process of radiative cooling is performed when the heat flux generated by an object on the outer surface of the infrared spectral layer within, 8–14 μm thick, is greater than the heat flux entering the layer by radiation, convection, or transfer (Wang et al., 2020). The radiative cooling in the PV modules increases when there is a spectrum shift in the surface emissivity of the modules for both thermal radiation and absorption (Ghadikolaei et al., 2021).

3.2. Energy Analysis in Radiative Cooling

The energy balance of radiative cooling on the PV panel can be analyzed as shown in Figure 7. The net cooling power, P_{cool} of such a radiative cooler is given by (Bijarniya et al., 2020).

$$P_{cool}(T) = P_{rad}(T) - P_{atm}(T_{amb}) - P_{sun} - P_{cond+conv}\,(T_s + T_{amb}) \tag{2}$$

$$P_{rad}(T_s) = 2\pi \int_0^{\frac{\pi}{2}} sin\theta cos\theta \int_0^{\infty} I_B(T_{s,\lambda})\varepsilon(\lambda,\theta)d\theta d\lambda \tag{3}$$

$$P_{atm}(T_{amb}) = 2\pi \int_0^{\frac{\pi}{2}} sin\theta cos\theta \int_0^{\infty} I_B(T_{amb,\lambda})\varepsilon(\lambda,\theta)\varepsilon_{atm}(\lambda,\theta)d\theta d\lambda \tag{4}$$

$$I_B(T,\lambda) = \frac{2hc^2}{\lambda^5(e^{\frac{hc}{\lambda k}}{}^{-1}} \quad (5)$$

$$\varepsilon_{atm}(\lambda,\theta) = 1 - t(\lambda)^{\frac{1}{cos\theta}} \quad (6)$$

$$P_{solar} = \int_0^{\infty} I_{solar}(\lambda)\varepsilon(\lambda,\theta_{solar})d\lambda \quad (7)$$

$$P_{cond+conv}(T_s + T_{amb}) = h_c(T_{amb} - T_s) \quad (8)$$

where $\varepsilon(\lambda, \theta)$ is the spectral and angular emissivity of a radiative cooler, atmospheric transmittance is $t(\lambda)$, and $\varepsilon_{atm}(\lambda, \theta)$ is atmospheric emissivity, which depends upon wavelength and zenith (θ), h_c combined convection and conduction heat transfer coefficient. Emissivity calculation is based on reflectivity for opaque surfaces (Raman et al., 2014).

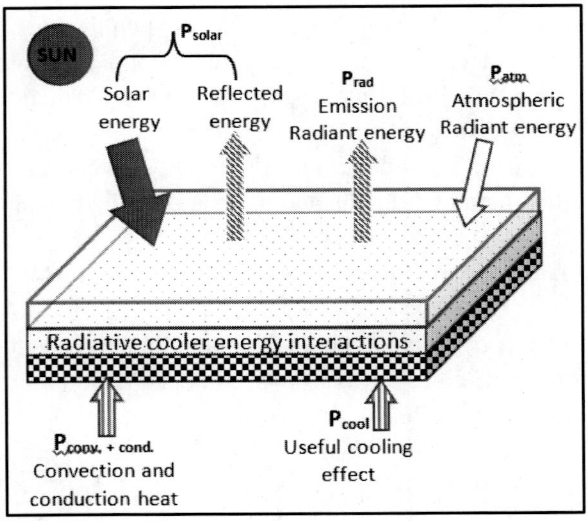

Figure 7. Energy balance of passive radiative cooling.

3.3. Review and Analysis of Radiative Cooling

Zhu et al., 2015 systematically proposed and demonstrated the concept of radiative cooling of solar cells. He designed a transparent cooler consisting of a two-dimensional square network of silica pyramids and a uniform silica layer with a thickness of 100 micrometers, as shown in Figure 8. This coolant is optically transparent and emits as a blackbody in the mid-infrared wavelength range. The results of performance predictions showed that even at a solar heating power of 800 W/m^2, the temperature reduction of silicon solar cells with this hierarchical cooler design can reach 17.6°C, which subsequently leads to an improvement in the relative electrical efficiency of 7.9%.

A transparent cooler was designed and manufactured by Zhu et al., 2015 as shown in Figure 9a, by drilling grooves for air holes with a depth of 500 micrometers from a double-sided polished silica wafer. Spectral characterization showed that this groove cooler has excellent heat emissivity in the infrared wavelength range and high sun transmittance (as shown in Figure 9b. To confirm the effect of radiative cooling, comparative experiments, as shown in Figure 9c, were presented to isolate the effects of ambient winds on the solar cells. The results showed that the temperature of the silicon solar cells can decrease negatively by 13 degrees Celsius.

However, Zhou et al., 2017 designed the soda-lime porous glass cooler so that the porous coating enhances the emission of mid-infrared radiation, allowing radiative cooling under direct sunlight. Moreover, the proposed cooler is mostly transparent. It has a high solar energy transmission ratio, and when using this cooler, the temperature of the PV module was reduced to 308 K (about 15 K higher than the ambient temperature). It is noted that it was much smaller when comparing this cooler with a traditional copper cooler. In addition, the efficiency of the lime glass cooler is about 3.7 times higher than the rest of the coolers' conventional refrigerants.

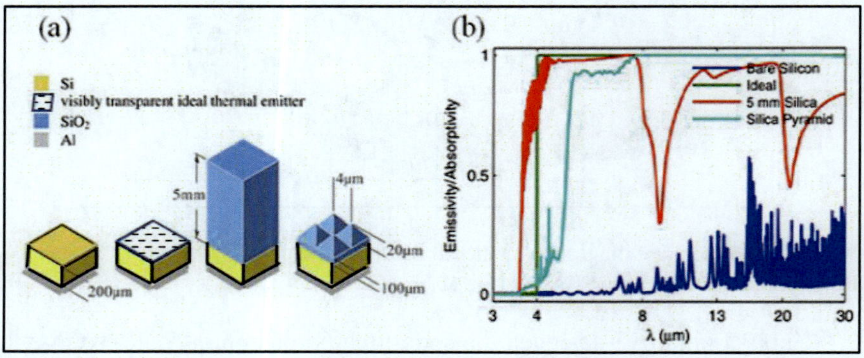

Figure 8. (a) Schematic of pyramid cooler and(b) the corresponding emissivity over the mid-infrared wavelength band (Zhu et al., 2015).

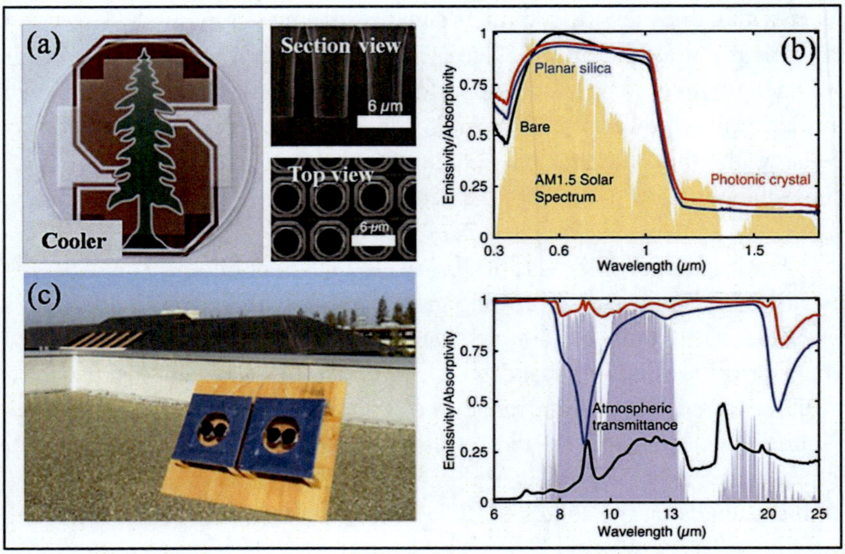

Figure 9. (a) Schematic of an air hole grooves silica cooler design (b) Measured spectral designed cooler (c) The outdoor experiment apparatus (Zhu et al., 2015).

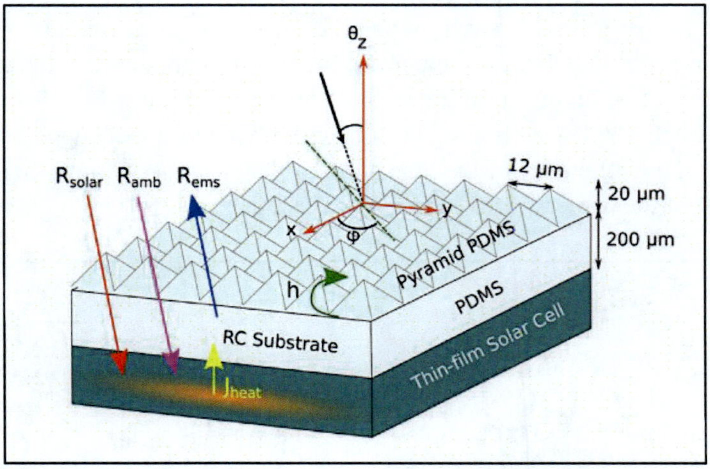

Figure 10. Schematic diagram of a hierarchical coolant made of polydimethylsiloxane (PDMS) (Lee & Luo, 2019).

Zhou et al., 2017 designed the soda-lime porous glass cooler so that the porous coating enhances the emission of mid-infrared radiation, allowing radiative cooling under direct sunlight. Moreover, the proposed cooler is mostly transparent. It has a high solar energy transmission ratio, and when using this cooler, the temperature of the PV module was reduced to 308 K (about 15 K higher than the ambient temperature). It is noted that it was much smaller when comparing this cooler with a traditional copper cooler. In addition, the efficiency of the lime glass cooler is about 3.7 times higher than the rest of the coolers' conventional refrigerants.

Moreover, hierarchical polydimethylsiloxane (PDMS) coolant was proposed by (Lee & Luo, 2019), as shown in Figure 10, where thermal analysis showed that the hierarchical PDMS layer could reduce the temperature of the solar cell by about 11°C – 16°C.

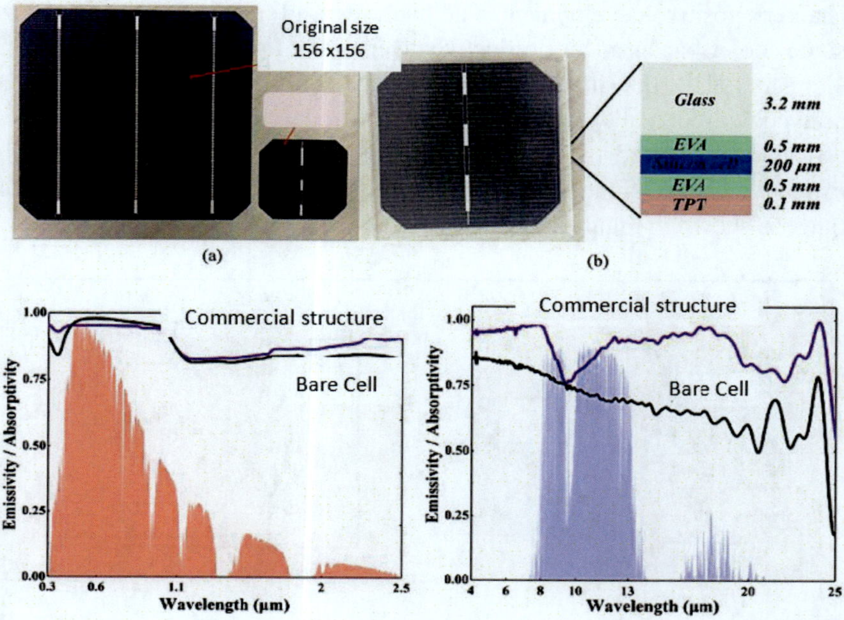

Figure 11. Schematic of encapsulated silicon solar cell with glass and a modified structure, where the latter has an additional PDMS layer to enhance the thermal emissivity and measured spectral property of the two samples (Zhao et al., 2018).

To elucidate the effect of a glass envelope on the radiative cooling property of enclosed solar cells, (Zhao et al., 2018) placed a thin layer of PDMS on a bare silicon cell. Then, the same layer was placed on top of a silicon cell coated with a cover made of glass to enhance the heat emission process, as shown in Figure 11. It shows the results of the test that was conducted in the environment. Normally, the temperature of both structures remained constant throughout the evaluation period. This provides evidence that there is a limit to the amount of radiative cooling capacity that can be enhanced by glass-enclosed solar cells. Altering the radiative property of the cover glass is one of several things that several researchers are still working on to increase solar cells' radiative cooling efficiency. In the past few years, this has been one of the topics that has garnered much attention.

In addition to the already improved ability for radiative cooling, another interesting finding related to this topic is that bare solar cells' sun absorption can be improved by placing specially designed coolers on top of the cells. This finding is related to the fact that the ability to radiate cooling has already been improved. In addition, the radiant cooling capacity has been improved, which is a very positive development. The main reason for this phenomenon is that some coolers act as an anti-reflective layer and/or light-trapping structure.

An et al., 2019 simulated multidimensional models are multilayer films arrayed with SiO_2 above a Si solar cell. 10°C reduced the temperature, the cell efficiency increased by 0.43%, and SiO_2 was chosen because of its high extinction coefficient in the infrared and high range. Transmission in the solar range is shown in Figure 12.

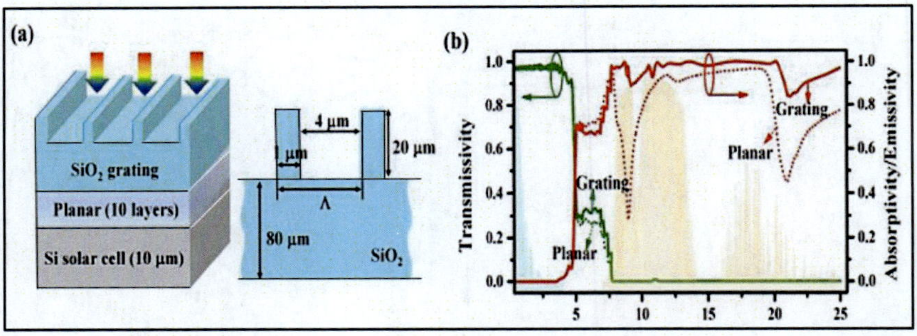

Figure 12. (a) Schematic of radiative quenching system applied in c-Si SC and integrates a multilayer thin film stack and SiO_2 grating. (b) Transmittance and emissivity simulations of planar and grating silica structures (An et al., 2019).

Zhou et al., 2016 designed a cooled structure of low-iron soda-lime glass. The proposed structure contains a two-dimensional square lattice of air holes etched into a 1.5-mm-thick low-iron glass sheet. The radius and depth of each antenna hole are 2.5 µm and 10 µm, respectively, as shown in Figure 13. This was done in order to improve the heat emission of the glass and thus increase the cooling capacity of the glass cover by radiative cooling, which was able to raise the efficiency of the photovoltaic panel due to cooling to 18%.

An experimental study of a photovoltaic water filter used by Al-Shohani et al., 2016 for PVT and concentrated photovoltaic/thermal (CPVT) systems. The optical effect of the water layer was studied to use the water layer in front of the photovoltaic cells instead of behind them when the photovoltaic module is combined either with the thermal system only (PVT) or with both the thermal and condenser systems (CPVT). Practically speaking, this study determines the transmittance efficiency of a water filter within the wavelength range (0.35–1 μm), which is the effective spectrum range of photovoltaic cells. As a result, the effects of the water layer thickness (1–5 cm) of the optical water filter were determined, as shown in Figure 24.

The results showed that the thickness of the water layer has a significant effect on the permeability efficiency, as the permeability decreases with the increase of the water layer. In general, the water layer over the photovoltaic cells in the PVT and CPVT module has positive advantages. It is able to enhance the photovoltaic thermal performance, as well as prevent cyclic thermal stresses on the photovoltaic cells.

All cooling techniques are thought to be workable options, but they have certain unique qualities. A comparison of conventional and radiative cooling techniques is shown in Table 3.

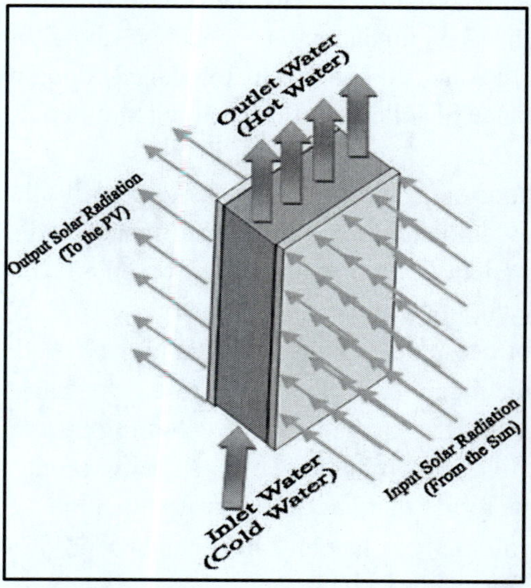

Figure 24. Water filter (water layer placed between two panels of glass subjected to solar radiation) (Al-Shohani et al., 2016).

Table 3. Comparison between Radiative cooling and other traditional cooling systems

Traditional cooling systems	Radiative cooling
• There are moving parts, such as pumps and tubes. • Forced cooling is applied. • Active cooling systems require electrical energy, either supplied from outside or generated by the system itself, in order to drive fans and pumps to remove heat from cell surfaces. • It needs a space that may be large, which may hinder the process of installing the panels perfectly. • It is more expensive than radiative cooling techniques. • Higher efficiency.	• There are neither moving parts nor external components required during the cooling process, which makes it very suitable for implementation on an industrial scale. • It does not need electrical energy to carry out its primary function of maintaining a cool environment for the solar panels. • It does not need much space; it is a thin film or a transparent cooler of small thickness. • Less cost in the long run. • Lower efficiency.

Conclusion

Within the context of this chapter, recent advancements in spectral selectivity methods and radiative cooling of solar cells are examined, studied, debated, and summarized. The fundamental concepts underlying the spectrally selective method of solar cell cooling are explained, along with an explanation of the energy balance of solar cells. The following outcomes are noteworthy:

- For outdoor optoelectronic systems, especially those with high heat loads and temperature-sensitive performance, radiative cooling might be a crucial thermal management method. In many situations, radiative cooling can be easily implemented. Due to its simplicity, a variety of radiative cooling materials may be used.
- One of the many things that many researchers are still working on to improve the efficiency of radiative cooling of solar cells is changing the radiative property of the cover glass. In recent years, this has been one of the topics that has drawn much attention.
- By controlling the solar cell's light and heat production, full-spectrum thermal management represents a revolutionary method for passively cooling solar cells.

- The key element of new refrigeration technology is a filter that is reliable and can be produced on a wide scale at an economically reasonable price.
- Radial cooling is less practical and less significant than cooling with full-spectrum thermal management technology.
- Radiation and spectrum-selective cooling techniques are still in their infancy on a practical and industrial level, but they will soon be crucial.

References

Agroui, K. (2012). Indoor and Outdoor Characterizations of Photovoltaic Module Based on Mulicrystalline Solar Cells. *Energy Procedia,* 18: 857-866. https://doi.org/10.1016/j.egypro.2012.05.100.

Agyekum, E. B., Kumar, S. P., Alwan, N. T., Velkin, V. I., Shcheklein, S. E., and Yaqoob, S. J. (2021). Experimental Investigation of the Effect of a Combination of Active and Passive Cooling Mechanism on the Thermal Characteristics and Efficiency of Solar PV Module. *Inventions*, 6(4): 63. https://doi.org/10.3390/inventions6040063.

Ahmed, S., Li, Z., Javed, M. S., & Ma, T. (2021). A Review on The Integration of Radiative Cooling and Solar Energy Harvesting. *Materials Today Energy*, 21: 100776. https://doi.org/10.1016/j.mtener.2021.100776.

Al-Ghezi, M.K.S., Ahmed, R.T., & Chaichan, M.T. (2022). The Influence of Temperature and Irradiance on Performance of the Photovoltaic Panel in the Middle of Iraq. *International Journal of Renewable Energy Development*, 11(2): 501-513. https://doi.org/10.14710/ijred.2022.43713.

Al-Shohani, W. A. M., Sabouri, A., Al-Dadah, R., Mahmoud, S. and Butt, H. (2016). Experimental Investigation of an Optical Water Filter for Photovoltaic/Thermal Conversion Module. *Energy Conversion and Management*, 111: 431-442. https://doi.org/10.1016/j.enconman.2015.12.065.

An, Y., Sheng, C., & Li, X. (2019). Radiative cooling of solar cells: Opto-Electro-Thermal Physics and Modeling. *Nanoscale*, 11(36): 17073-17083. https://doi.org/10.1039/c9nr04110a.

Bijarniya, J. P., Sarkar, J., & Maiti, P. (2020). Review on Passive Daytime Radiative Cooling: Fundamentals, Recent Researches, Challenges and Opportunities. *Renewable and Sustainable Energy Reviews*, 133: 110263. https://doi.org/10.1016/j.rser.2020.110263.

Dubey, S., Sarvaiya, J. N., & Seshadri, B. (2013). Temperature Dependent Photovoltaic (PV) Efficiency and Its Effect on PV Production in the World - A Review. *Energy Procedia*, 33: 311-321. https://doi.org/10.1016/j.egypro.2013.05.072.

Dwivedi, P., Sudhakar, K., Soni, A., Solomin, E., I Kirpichnikova, I. (2020). Advanced Cooling Techniques of PV Modules: A State Of Art. *Case Studies in Thermal Engineering*, 21: 100674. https://doi.org/10.1016/j.csite.2020.100674.

Gentle, A. R., & Smith, G. B. (2016). Is Enhanced Radiative Cooling of Solar Cell Modules Worth Pursuing? *Solar Energy Materials and Solar Cells*, 150: 39-42. https://doi.org/10.1016/j.solmat.2016.01.039.

Gu, W., Ma, T., Ahmed, S., Zhang, Y., & Peng, J. (2020). A Comprehensive Review and Outlook of Bifacial Photovoltaic (Bpv) Technology. *Energy Conversion and Management*, 223. https://doi.org/10.1016/j.enconman.2020.113283.

Homadi, A., Hall, T., & Whitman, L. (2020). Study a Novel Hybrid System for Cooling Solar Panels and Generate Power. *Applied Thermal Engineering* 179, 115503. https://doi.org/10.1016/j.applthermaleng.2020.115503.

Kalaiselvan, S., Karthikeyan, V., Rajesh, G., Kumaran, A.S., Ramkiran, B., and Neelamegam, P. (2018). Solar PV Active and Passive Cooling Technologies-A Review. *7th IEEE International Conference on Computation of Power, Energy, Information and Communication, ICCPEIC 2018*, 166-169. https://doi.org/10.1109/ICCPEIC.2018.8525185.

Koohestani, S. S., Nižetić, S., & Santamouris, M. (2023). Comparative Review and Evaluation of State-Of-The-Art Photovoltaic Cooling Technologies. *Journal of Cleaner Production*, 136953. https://doi.org/10.1016/j.jclepro.2023.136953.

Krotkus, A. Nevinskas, I. Norkus, R. Geižutis, A., Strazdienė, V., Pačebutas, V., and Paulauskas, T. (2023). Terahertz Photocurrent Spectrum Analysis of Algaas/Gaas/Gaasbi Multi-Junction Solar Cells. *Journal of Physics D: Applied Physics*, 56(35). https://doi.org/10.1088/1361-6463/acd85d.

Lee, E., & Luo, T. (2019). Black Body-Like Radiative Cooling for Flexible Thin-Film Solar Cells. *Solar Energy Materials and Solar Cells*, 194: 222-228. https://doi.org/10.1016/j.solmat.2019.02.015.

Li, W., Shi, Y., Chen, K., Zhu, L., and Fan, S. (2017). A Comprehensive Photonic Approach for Solar Cell Cooling. *ACS Photonics*, 4(4): 774-782. https://doi.org/10.1021/acsphotonics.7b00089.

Li, Z., Ahmed, S., & Ma, T. (2021). Investigating the Effect of Radiative Cooling on the Operating Temperature of Photovoltaic Modules. *Solar RRL*, 5(4): 1-10. https://doi.org/10.1002/solr.202000735.

Long, L., Yang, Y., & Wang, L. (2019). Simultaneously Enhanced Solar Absorption and Radiative Cooling With Thin Silica Micro-Grating Coatings for Silicon Solar Cells. *Solar Energy Materials and Solar Cells*, 197: 19-24. https://doi.org/10.1016/j.solmat.2019.04.006.

Lu, K., Zhao, B., Xu, C., Li, X. and Pei, G. (2022). A full-Spectrum *Synergetic* Management Strategy for Passive Cooling of Solar Cells. *Solar Energy Materials and Solar Cells*, 245: 111860. https://doi.org/10.1016/j.solmat.2022.111860.

Lu, Y., Chen, Z., Ai, L., Zhang, X., Zhang, J., Li, J., Wang, W., Tan, R., Dai, N., and Song, W. (2017). A Universal Route to Realize Radiative Cooling and Light Management in Photovoltaic Modules. *Solar RRL*, 1(10): 1-7. https://doi.org/10.1002/solr.201700084.

Lupu, A. G., Homutescu, V. M., Balanescu, D. T., & Popescu, A. (2018). A Review of Solar Photovoltaic Systems Cooling Technologies. *IOP Conference Series: Materials Science and Engineering*, 444(8). https://doi.org/10.1088/1757-899X/444/8/082016.

Ghadikolaei, S. S. C. (2021). Solar Photovoltaic Cells Performance Improvement by Cooling Technology: An Overall Review. *International Journal of Hydrogen Energy*, 46(18): 10939-10972. https://doi.org/10.1016/j.ijhydene.2020.12.164.

Mahdavi, A., Farhadi, M., Gorji-Bandpy, M., & Mahmoudi, A. (2022). A Review of Passive Cooling of Photovoltaic Devices. *Cleaner Engineering and Technology*, 11: 100579. https://doi.org/10.1016/j.clet.2022.100579.

Mandal, J., Fu, Y., Overvig, A.C., Jia, M., Sun, K., Shi, N., Zhou, H., Xiao, X., Yu, N. and Yuan Yang (2018). Hierarchically Porous Polymer Coatings for Highly Efficient Passive Daytime Radiative Cooling. *Science*, 362(6412): 315-319. https://doi.org/10.1126/science.aat9513.

Olita, F. (2012). *Advanced Control and Condition Monitoring PV Systems*. Student project, Aalborg University. Norway.

Perrakis, G., Tasolamprou, A. C., Kenanakis, G., Economou, E. N., Tzortzakis, S. and Kafesaki, M. (2020). Passive Radiative Cooling And Other Photonic Approaches For The Temperature Control of Photovoltaics: A Comparative Study for Crystalline Silicon-Based Architectures. *Optics Express*, 28(13): 18548. https://doi.org/10.1364/oe.388208.

Radziemska, E. (2003). The Effect of Temperature Difference on Heat Transport in Solar Cells. *Archives of Thermodynamics*, 24(3): 83-95.

Raman, A. P., Anoma, M. A., Zhu, L., Rephaeli, E., and Shanhui Fan, S. (2014). Passive Radiative Cooling Below Ambient Air Temperature Under Direct Sunlight. *Nature*, 515(7528), 540-544. https://doi.org/10.1038/nature13883.

Raval, H. D., Maiti, S., & Mittal, A. (2014). Computational Fluid Dynamics Analysis and Experimental Validation of Improvement in Overall Energy Efficiency of a Solar Photovoltaic Panel By Thermal Energy Recovery. *Journal of Renewable and Sustainable Energy*, 6(3). https://doi.org/10.1063/1.4885178.

Rephaeli, E., Raman, A., & Fan, S. (2013). Ultrabroadband Photonic Structures to Achieve High-Performance Daytime Radiative Cooling. *Nano Letters*, 13(4): 1457-1461. https://doi.org/10.1021/nl4004283.

Safi, T. S., & Munday, J. N. (2015). Improving Photovoltaic Performance Through Radiative Cooling in Both Terrestrial and Extraterrestrial Environments. *Optics Express*, 23(19), A1120. https://doi.org/10.1364/oe.23.0a1120.

Sato, D., & Yamada, N. (2019). Review of Photovoltaic Module Cooling Methods and Performance Evaluation of The Radiative Cooling Method. *Renewable and Sustainable Energy Reviews*, 104: 151-166. https://doi.org/10.1016/j.rser.2018.12.051.

Slauch, I. M., Deceglie, M. G., Silverman, T. J., & Ferry, V. E. (2018). Spectrally Selective Mirrors with Combined Optical and Thermal Benefit for Photovoltaic Module Thermal Management. *ACS Photonics*, 5(4): 1528-1538. https://doi.org/10.1021/acsphotonics.7b01586.

Sun, X., Silverman, T. J., Zhou, Z., Khan, M. R., Berme, P., and Alam, M. A. (2017). Optics-Based Approach to Thermal Management of Photovoltaics: Selective-Spectral and Radiative Cooling. *IEEE Journal of Photovoltaics*, 7(2): 566-574. https://doi.org/10.1109/JPHOTOV.2016.2646062.

Sun, Y., Zhou, Z., Jin, X., Sun, X., Alam, M. A., and Bermel, P. (2017). Radiative cooling for concentrating photovoltaic systems. In *Proceedings SPIE, Thermal Radiation Management for Energy Applications, 103690D* (6 September), San Diego, California, United States. https://doi.org/10.1117/12.2273916.

Wang, Z., Kortge, D., Zhu, J. Zhou, Z., Torsina, H., Lee, C., and Bermel, P. (2020). Article Lightweight, Passive Radiative Cooling to Enhance Concentrating Photovoltaics Lightweight, Passive Radiative Cooling to Enhance Concentrating Photovoltaics. *Joule*, 4(12): 2702-2717. https://doi.org/10.1016/j.joule.2020.10.004.

Zhao, B., Hu, M., Ao, X., & Pei, G. (2018). Performance Analysis of Enhanced Radiative Cooling Of Solar Cells Based on A Commercial Silicon Photovoltaic Module. *Solar Energy*, 176: 248-255. https://doi.org/10.1016/j.solener.2018.10.043.

Zhao, B., Hu, M., Ao, X., Xuan, Q., and Pei, G. (2018). Comprehensive Photonic Approach for Diurnal Photovoltaic and Nocturnal Radiative Cooling. *Solar Energy Materials and Solar Cells*, 178: 266-272. https://doi.org/10.1016/j.solmat.2018.01.023.

Zhao, B. Hu, M., Ao, X., Xuan, Q., Pei, G. (2020). Spectrally Selective Approaches For Passive Cooling of Solar Cells: A Review. *Applied Energy*, 262: 114548. https://doi.org/10.1016/j.apenergy.2020.114548.

Zhao, B., Lu, K., Hu, M., Liu, J., Wu, L., Xu, C., Xuan, Q., Pei, G. (2022). Radiative Cooling of Solar Cells With Micro-Grating Photonic Cooler. *Renewable Energy*, 191: 662–668. https://doi.org/10.1016/j.renene.2022.04.063.

Zhao, D Aili, A., Zhai, Y., Xu; S., Tan, G., Yin, X., and Yang, R. (2019). Radiative Sky Cooling: Fundamental Principles, Materials, and Applications. *Applied Physics Reviews*, 6(2). https://doi.org/10.1063/1.5087281.

Zhou, Z., Sun, X., & Bermel, P. (2016). Radiative Cooling for Thermophotovoltaic Systems. *Infrared Remote Sensing and Instrumentation XXIV*, 9973: 997308. https://doi.org/10.1117/12.2236174.

Zhou, Z., Wang, Z., and Bermel, P. (2019). Radiative Cooling for Low-Bandgap Photovoltaics Under Concentrated Sunlight. *Optics Express*, 27(8): A404. https://doi.org/10.1364/oe.27.00a404.

Zhu, L., Raman, A. P., and Fan, S. (2015). Radiative Cooling of Solar Absorbers Using a Visibly Transparent Photonic Crystal Thermal Blackbody. In *Proceedings of the National Academy of Sciences of the United States of America*, 112(40): 12282-12287. https://doi.org/10.1073/pnas.1509453112.

Zhu, L., Raman, A., Wang, K. X., Anoma, M. A., and Fan, S. (2015). Radiative Cooling for Solar Cells. *Physics, Simulation, and Photonic Engineering of Photovoltaic Devices IV*, 9358: 935814. https://doi.org/10.1117/12.2080148.

Chapter 8

Productivity Enhancement of Hybrid Double Slope Solar Still for Desalination

Mariam E. Murad[1,*]
Wissam H. Alawee[2,3]
Hayder A. Dhahad[1]
and Zakaria. M. Omara[4]
[1]Mechanical Engineering Department, University of Technology- Iraq, Baghdad, Iraq
[2]Control and Systems Engineering Department, University of Technology- Iraq, Baghdad, Iraq
[3]Training and Workshops Center, University of Technology- Iraq, Baghdad, Iraq
[4]Mechanical Engineering Department, Faculty of Engineering, Kafrelsheikh University, Kafrelsheikh, Egypt

Abstract

The global scarcity of fresh water has prompted the need for sustainable solutions. Solar distillation, utilizing solar energy to generate clean water, is effective for communities facing inadequate freshwater sources. This chapter evaluates the effectiveness of a modified double-slope solar still (MDSSS) compared to conventional double-slope solar still. The MDSSS incorporates a solar collector and phase change materials. The results demonstrate that implementing energy storage materials inside solar stills significantly enhances distillate output, potentially improving it by up to 42%. Additionally, modifications to solar stills, such as finned corrugated basins, heat storage materials, wick materials, and external modifications like reflectors, can lead to energy efficiency improvements of up to 30.42%. Active solar stills integrated with flat-plate solar

* Corresponding Author's Email: me.21.14@grad.uotchnology.edu.iq.

In: Advancements in Sustainability Systems
Editors: Hussain H. Al-Kayiem, Hasanain A. Abdul Wahhab and Raed A, Jessam
ISBN: 979-8-89113-571-0

collectors can achieve remarkable productivity rates of up to particularly when employing cover cooling flash tactics with alternating 5-minute intervals on and off. The chapter provides valuable guidance for future advancements in solar desalination technologies. These advancements contribute to developing more efficient and sustainable water production systems, ensuring a reliable and accessible clean water supply for communities facing water scarcity challenges.

Keywords: clean water, double slope solar still, desalination, solar thermal energy, water scarcity

1. Introduction

Water is necessary for the life sustainability of humans, animals, plants, and other creatures. The availability of clean water is an essential sustainable goal for human activities, agriculture, industry, and nature balance. In areas with difficult access to clean water resources, solar desalination has proved to be an effective means to produce limited amounts of clean water. The typical processes of distillation by reverse osmosis and electrodialysis for saltwater desalinating to produce fresh water need much energy. On the other hand, desalination only needs to be done in a few cubic meters per day in remote, dry areas.

Solar stills have emerged as a promising solution, harnessing solar energy to purify water through evaporation and condensation (Al-Kayiem, 2019; Hammoodi et al., 2023). Among these, the double-slope solar still has garnered attention for its potential to enhance productivity and efficiency. This design has long been recognized as a valuable technology for water purification and desalination in areas with limited access to fresh water. Utilizing solar energy facilitates the evaporation of water, the separation of impurities, and the collection of purified condensed vapor. The double-slope solar still offers several advantages over other designs, including improved productivity and efficiency (Al-Kayiem et al., 2023; Alawee et al., 2021; Pandey, 1983).

Several studies aimed to improve the effectiveness of solar distillation systems through various modifications, such as altering the distiller's shape and incorporating additional components to increase water temperature (Selvaraj & Natarajan, 2018; Sharshir et al., 2020). The temperature of the glass cover can be successfully reduced by implementing certain measures, such as the use of continuously running water or intermittent cooling water on the lid, as reported in previous studies (Gupta et al., 2017; Janarthanan et al., 2006). Additionally, it has been observed that wind velocity can impact the surface temperature and subsequently enhance the production rate of the still, as noted in another study (Alawee et al., 2021; Kabeel et al., 2017). Efficiency has been improved by developing solar stills with solar collectors, condensers, heat storage, and heat recycling, as well as color multi-stage/multi-effect solar stills and hybrid solar still/PV systems (Essa, 2022; Hammoodi et al., 2023). Because they do not boil the water, solar distillers cannot eliminate bacteria or degrade harmful chemicals. Also, insects and other pests may congregate in the space designated for the slanted glass cover, lowering the quality and amount of water that can be generated.

However, solar stills commercial feasibility remains restricted due to limited production. Research has been conducted on utilizing phase change materials (PCM) possessing high latent heat capacity for energy absorption and productivity enhancement, as evidenced by sources (Agyenim et al., 2010; El-Sebaii, Al-Ghamdi et al., 2009). In recent years, scholars and professionals have been investigating diverse techniques to enhance the efficiency of double-slope solar stills (Rajaseenivasan et al., 2015; Tuly et al., 2021). An approach that shows promise involves the integration of fins, collectors, and thermal energy storage (TES) materials into the design. This integration aims to optimize the process of heat transfer, augment energy absorption, and enhance overall performance (Alawee, 2015).

While double-slope solar stills have been previously recognized as effective water purification and desalination technologies, incorporating these elements represents a novel approach to enhancing their productivity and efficiency. Integration of the solar stills with collectors' aids in acquiring and consolidating solar radiation, thereby optimizing the quantity of energy accessible for the distillation procedure. Through the more efficient concentration of sunlight on the surface of the still, collectors can augment the temperature gradient, thereby enhancing the evaporation rate (Fathy et al., 2018). Moreover, the incorporation of TES substances facilitates the capacity of the double-slope solar still to accumulate surplus heat throughout maximum sunlight exposure. The materials can assimilate and retain thermal energy, thereby enabling their sustained operation during periods of diminished solar radiation or nocturnal hours. The gradual release of stored heat enables the maintenance of a stable temperature within the system, thereby extending the operational duration of the still (Nayi & Modi, 2018). Integrating collectors and TES materials is crucial in enhancing the productivity of double-slope solar stills. The utilization of this technology exhibits significant promise in ameliorating water scarcity and catering to the escalating need for uncontaminated and readily available water resources in arid areas. Through the optimization of the distillation process and the maximization of solar energy utilization, this innovation has the potential to make a significant contribution to sustainable water management and alleviate water stress in diverse communities (Tuly et al., 2021). A comprehensive review and comparison between various hybrid solar stills are performed by Al-Kayiem et al., 2023 including performance economic analysis.

This chapter investigates the importance of incorporating collectors and thermal storage materials into double-slope solar stills. The present study delves into the advantages and obstacles of this methodology, scrutinizes the most recent scientific discoveries and technological progressions, and deliberates on the possibility of extensive adoption. Through comprehension and utilization of this technology, it is plausible to establish a more proficient and enduring approach toward water purification and desalination.

2. Fundamentals of a Solar Still

Solar stills, as solar systems for desalination, are harnessing solar energy for water evaporation-condensation processes. The inlet water supply may be in a batch or at a low continuous flow rate.

2.1. Components of Solar Stills

Typically, solar stills consist of the following components (Al-Kayiem et al., 2023):

- Basin or Collection Chamber: This is the lower part of the solar still where the impure water is placed for purification. It can be a shallow basin, or a dug-out area lined with a waterproof material to prevent leakage.
- Transparent Cover: The transparent cover is placed over the basin or collection chamber to allow sunlight to enter and heat the water. It is usually made of glass or plastic and acts as a barrier to prevent the escape of water vapor.

- Condensation Surface: The inner surface of the transparent cover serves as the condensation surface. As the water vapor rises and encounters the cooler surface, it condenses, and forms purified water droplets.
- Collection System: A collection channel or collection basin is placed below the condensation surface to collect the condensed water droplets. It allows the purified water to flow and accumulate for further use.
- Insulation (optional): In some solar still designs, insulation materials such as foam or polystyrene may be added to the sides or bottom of the basin to reduce heat loss and improve efficiency.
- Support Structure: A framework or support structure holds the transparent cover in place and stabilizes the solar still.
- Sealing Materials: To ensure proper functioning and prevent leakage, sealing materials such as silicone or rubber gaskets may be used to seal the edges of the transparent cover and the basin.

However, different designs may incorporate additional features or modifications based on specific requirements and conditions.

2.2. Operational Principles of Solar Stills

The transparent glass or plastic cover facilitates sunlight entry in this solar still design, which heats the contaminated water in the underlying basin or pan. The water's temperature increases results in its conversion into water vapor confined within the still. The warm and humid air ascends towards the inclined glass surface and undergoes condensation on the lower side. This phenomenon occurs due to the cooling effect of the ambient air outside. Consequently, impurities such as salts, minerals, and other pollutants are deposited in the pan.

When the transparent cover made of glass or plastic is inclined at an angle of approximately 10 to 20 degrees, the condensed water droplets that emerge on the underside of the cover coalesce, forming larger water droplets. These droplets increase in weight as they grow and roll down the inclined glass cover into a collector tube or trough at the bottom, facilitating the distillate's drainage. Distilled water, which is clean and condensed, may be gathered in a pristine receptacle for preservation. The procedure yields purified water that is free of any impurities and can be used for diverse purposes. Any particulate matter or pollutants are confined within the still and necessitate periodic elimination. Figure 1 depicts the procedure for a double-sloped, simple solar still.

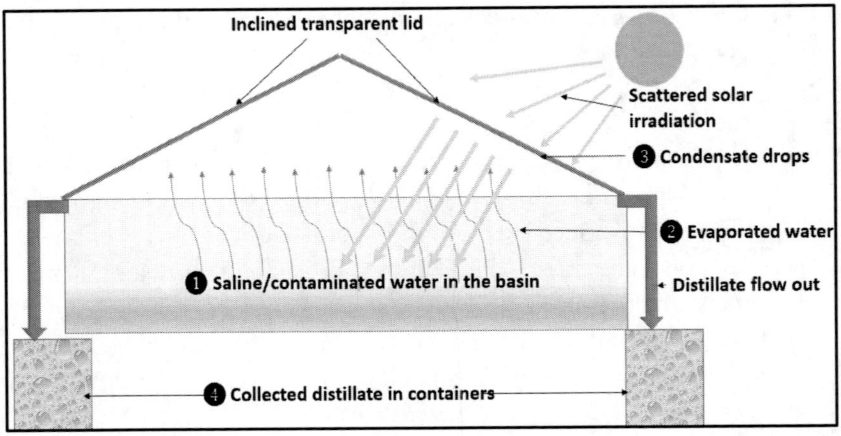

Figure 1. Working principle of a double-slope solar still.

3. Hybrid Double Slope Solar Still

A research approach to enhance the design of solar stills (SS) is the utilization of double-slope solar stills. The primary objective of this approach is to improve and optimize the freshwater production rate. Substantial research efforts have been dedicated to analyzing and evaluating the performance of double-slope solar stills. The focus of these studies is to investigate and understand the factors that impact the efficiency and productivity of double-slope solar stills. Researchers aim to identify strategies and design enhancements that can maximize solar energy utilization and optimize the processes of water evaporation and condensation.

Various aspects are examined in the research on double-slope solar stills, including the geometry and inclination angles of the slopes, the materials used, the design of the condensation surfaces, and the overall system dynamics. By thoroughly analyzing the performance of double-slope solar stills, researchers strive to develop innovative solutions and improvements that can significantly increase the freshwater production rate.

4. Literature Survey of Solar Distillation Systems

Previous research in enhancing the production of double-slope solar stills has focused on various aspects to improve their efficiency and productivity. These studies have provided valuable insights into different methods and techniques that can be employed to optimize the performance of solar stills. By examining and analyzing the findings of these previous studies, researchers have identified key areas for improvement and developed innovative approaches. This includes the investigation of novel design modifications, the incorporation of PCM and advanced materials, the investigation of heat transfer enhancement techniques, the consideration of integrated systems, and the application of advanced modeling and optimization.

4.1. Hybrid Double Slope Solar Still with PCMs

Agrawal and Singh, 2022 tested the performance of a conventional double-slope single-basin solar still and a modified double-slope single-basin solar still combined with binary eutectic PCM to determine their energy efficiency and hourly production. The atmospheric conditions of Bhopal, India (23.24° N, 77.40° E) are utilized to develop a multiphase, three-dimensional CFD model of the solar still in the ANSYS Fluent 19.2 workbench. Integrating PCM resulted in a notable enhancement in the average energy efficiency of modified solar still, elevating it from 22.21% to 30.42%. The modified's experimental daily efficiency was still found to be 30.42%, while the simulated value was determined to be 33.15%. The modified solar still is anticipated to produce a maximum freshwater yield of 0.31 kg/m^2 h at 3:00 PM, whereas the simulated outcome is 0.285 kg/m^2 h. The modified solar still exhibits a disparity of approximately 8.06% between its experimental and simulated outcomes. Likewise, the simulated outcome of conventional solar still exhibits an error of around 10.09% compared to the experimental data. Vapor leakage and heat transfer loss may explain why experimental values are lower than simulated values.

Moreover, Suneesh et al., 2013, designed and evaluated a solar still featuring a basin with a "V" shape, both with and without the incorporation of cement blocks. The experimental features of the samples are assessed and contrasted under comparable environmental circumstances. Experimental evaluation was conducted on the solar still during standard summer days in

Coimbatore (11°00 N, 77°00 E) under prevailing Indian climatic conditions. Based on the calculations, the system's efficiency remains at 24% when employing cement blocks, whereas it drops to 20% in their absence. The primary advantage of the still lies in the heat emission of its cement blocks, which occurs post 5:00 PM, thereby resulting in a 4% enhancement in efficiency. The high density of cement facilitates heat absorption during the peak hours of the day and its subsequent release during the night. Our analysis indicates that cement is a highly effective PCM for use in solar desalination. The planned design is expected to offer rural communities a viable approach to converting saline water into potable water. The utilization of solar still presents a viable solution for providing fresh water to rural and arid regions due to its uncomplicated nature.

Along with this, Patil and Dambal, 2016, conducted an empirical study in Karnataka, India, on a solar still with a double slope and a single basin. The study utilized PCMs like paraffin wax and sensible heat storage components like black pebbles. The solar still was fabricated using aluminum sheet metal and had a surface area of 0.7 m^2. The experiment was conducted in an open environment. A 0.40 m^2 aluminum tray was installed within the still, leaving a 100 mm gap. Thermocool was used for insulation between the spaces. Two glass covers with a thickness of 4mm and an area of 0.35 m^2 were placed on top of the still at an angle of 150 degrees, based on the latitude of Haliyal. In April, a performance analysis was carried out through three experiments involving a black-coated aluminum tray with PCM. The study utilized a Pyranometer and K-type thermocouple for analysis. The results showed that the percentage productivity was 30% for paraffin wax and black coated tray, 18% for black pebbles and black coated tray, and 13% for paraffin wax and black pebbles.

Murugavel et al., 2010 asserted that a single basin, double-slope solar still with energy-storing materials, including quartzite rock, red brick fragments, cement concrete fragments, washed stones, and iron scraps, was able to retain the extra energy and boost nighttime output. The most productive material in their studies was quartzite rock, which was 3/4 in size and produced 2.1 kg/m^2 of output, an increase of 6.2% above what it would have done with only the same quantity of water and no energy-storing material. Heat storage extends the working day overnight, increasing overall output. Although the installation and operation expenses would go up a bit, it is still advised for use in realistic applications (see Figure 2).

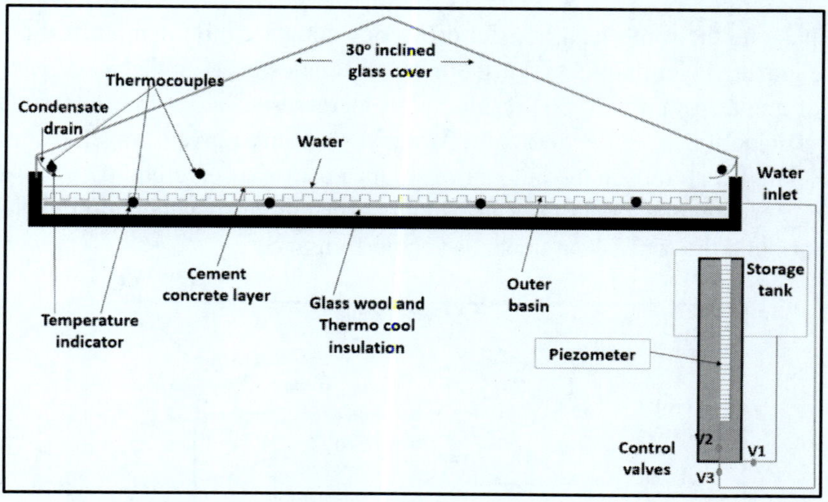

Figure 2. Single basin double slope solar still.

In a separate study In Nagar, Gujarat, India, during the first part of November 2013, Shah and Damor, 2013, examined the double-slope single basin theoretically and experimentally, with and without heat-absorbing materials. Heat-absorbing materials like rubber and black dye were employed on the basin's bottom side. The still was built with a base of 1 m by 1 m, and the glass cover was tilted at 35 degrees. The quantity of distilled water was monitored every hour, and the temperatures of the glass cover, the water within the still, and the surrounding air were all noted. A study on the impact of various heat-absorbing materials on the efficiency of the solar still reveals that the output of the still-using black dye as a heat-absorbing material is 26.7% higher than that without the use of absorbing materials, allowing to produce of a greater volume of distilled water.

Glazed roofs are commonly used in modern building designs because they provide better lighting. However, its less effective heat insulation might reduce building thermal comfort. To enhance the building's internal thermal climate and lower energy usage while obtaining enough interior lighting, PCM is injected into the interlayer of the insulating glazed roof, as shown in Figure 3. Mao and Yang, 2020, created a computational heat transfer model for a double-slope PCM-glazed roof. The heat transmission capabilities of double-slope hollow glazed roofs and double-slope PCM glazed roofs were compared. The findings indicate that the alteration of the phase change layer thickness in a double-slope PCM glazed roof has a greater impact on enhancing the roof's heat transfer efficiency compared to the modification of the glazing layer thickness. Furthermore, it can be observed that a decrease in roof angle from 45° to 30° results in an increase in the temperature control coefficient from -2.3% to 22.5%. Moreover, within a given range, a smaller roof angle can reduce the amount of solar radiation received by the roof, and a higher latent heat energy of PCM can absorb more solar radiation, thereby improving the roof's heat transmission performance.

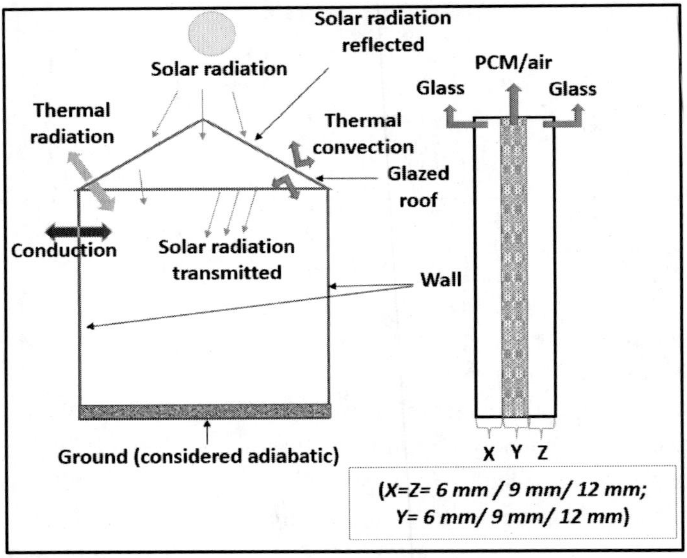

Figure 3. Heat transfer process of the double-slope PCM glazed roof.

Saravanan et al., 2021, studied how Kanchey marbles performed as an energy storage medium for single-basin and double-slope solar stills. The solar system still stores the sun's energy as sensible or latent heat. Sensible heat is only absorbed and used during the day because temperature differences do not cause phase changes, while latent heat is utilized only after sunset. It captures the excess solar energy that is accessible throughout the day and, by shifting its phase, provides it in the evening. The present study focuses mostly on improving still output performance utilizing various sensible heat storage materials, such as Kanchey marbles with a uniform size of 1. The yield of fresh water with Kanchey marbles was shown experimentally to be 16.32% greater than the yield without Kanchey marbles. The double-slope solar system, still utilizing marble as heat storage materials, produced the highest output of 4.094 l/m^2.

4.2. Hybrid Double Slope SS with PCMs and Condenser

By using PCMs as thermal storage material, Faegh and Shafii, 2017 developed the new concept of using solar stills to store the latent heat of condensing vapor. The water vapor produced by the solar energy throughout the day is sent to an external condenser that is filled with PCM to be condensed. PCM absorbs the lost latent heat and stores it as a result. The solar energy is not immediately stored in the PCM since there is no direct contact between the salty water and the PCM. The desalination process can continue in the evening thanks to the energy stored in the PCM being transmitted as heat to the salty water through heat pipes. A solar still with evacuated tube collectors: The findings indicate that incorporating an external condenser containing PCM and heat pipes into a solar still featuring evacuated tube collectors can sustain the desalination process beyond sunset while maintaining daytime yield levels. Incorporating PCM into the system resulted in an 86% increase in yield compared to the system without PCM. The yield reached 6.56 kg/m^2 per day with an efficiency of 50%. The current economic analysis indicates that the cost per liter per unit area of the produced water is 0.015 USD in the optimal case.

4.3. Hybrid Double Slope SS with PCM, Collector, and Fins

Prajapati et al., 2019 analyzed the effectiveness of a double-slope solar distiller by incorporating circular fins and PCM to increase potable water output. The modified solar distiller utilizes $Mg_2SO_4.7H_2O$ as the phase change material and incorporates round aluminum fins. A solar desalination system was evaluated by designing and constructing two double-slope solar stills with identical basin areas of 0.5 x 0.5 m^2. One of the solar stills under consideration is a conventional design that lacks fins or PCM, whereas the other incorporates aluminum fins and 1 kg of PCM. The empirical results demonstrate that the daily output of potable water from the solar distillation unit incorporating phase change material and aluminum fins surpasses that of the conventional solar still. On the initial day, the conventional solar stills and modified solar stills (with fins) yielded a daily freshwater productivity of roughly 1400 and 1700 ml/m^2 per day, respectively. On a subsequent day, the modified solar stills (with fins) and modified solar stills with 1.0 kg PCM at 5 cm water depth produced approximately 1560 and 1940 ml/m^2 per day of freshwater, respectively.

Besides that, Hedayati et al., 2020 conducted an experimental study on a double-slope solar still integrated with photovoltaic/thermal (PV/T) technology and utilizing paraffin wax as a PCM. The study's findings indicate that integrating PV/T technology increases basin water temperature during the daytime, resulting in increased production. Additionally, using PCM facilitates an increase in nocturnal production as it stores thermal energy during the day and releases it during the night. According to their findings, an increase in mass flow rate leads to a corresponding increase in exergy efficiency. The implementation of dual collectors enhanced exergy efficiency compared to using a single collector. The enhancement of exergy efficiency was observed with the augmentation in the number of collectors. However, the improvement above did not exhibit a substantial difference when the number of collectors exceeded two. Thus, considering the economic dimension and the escalation of expenses, two collectors were evaluated for the system. During the summer season, the highest amount of water production recorded was 6.5 kg/m^2.

4.4. Hybrid Double Slope SS with Heat Storage and External Reflector

Gnanaraj and Velmurugan, 2022, increased the distiller's output by distributing heat-storage material in the basin or adding an external reflector to direct more solar radiation into the still. Three identically sized solar stills were created for the research with the same proportions. They were the traditional distiller, the distiller equipped with heat storage, and the distiller equipped with an exterior reflector. The study aimed to determine how well single-basin and double-slope solar distillers performed when external reflectors and heat storage materials were used. The major goal was to study the rate of change of temperature parameter values and production values under rising and falling daylight hours. The rate of rise in water temperature and distillate output in the reflector-equipped distiller was rapid as the number of sunlight hours increased. However, the distiller with heat-storing materials maintained greater water temperatures and output rates throughout the dwindling daylight hours. The results demonstrate that the productivity of the heat storage material distiller was 23.08% higher than that of the traditional distiller. However, the distiller with a reflector had a production increase of 63.97%. Figure 4. depicts a solar still with an external reflector.

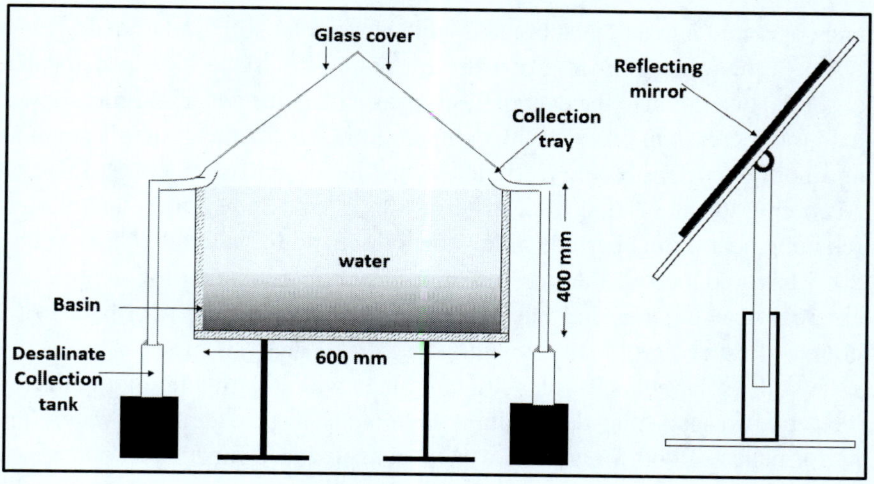

Figure 4. Double slope solar still with external reflector.

To increase the double-slope, single-basin solar still's productivity in producing distillate. Modifications were carried out both internally and externally to enhance the conventional design. Three solar stills with internal modifications were built, including ones with finned corrugated basins, black granite, and wicks. Another still was made, this time with reflectors attached to the outside (an external modification). Finally, one was built with all the modifications above already included. The conventional still produced 1880 ml/m^2.day of distillate. The distillate production of the still with the finned corrugated basin, the still with the black granite, the still with the wick, the still with the reflector, and the solar still with all internal and external modifications were 2995, 3210, 2690, 3655, and 5130 ml/m^2·day, respectively. These estimates were higher than the production of the conventional still by 58.47%, 69.84%, 42.33%, 93.39%, and 171.43%. The effect of the water-glass temperature differential and basin water temperature on production rate was also examined. Finally, theoretical values and experimental values were compared (Gnanaraj and Velmurugan, 2019).

4.5. Hybrid Double Slope SS with Various Types of Collector and Wick Materials

Al-Molhem and Eltawil, 2020, assessed the performance of a double-slope solar still that was combined with a flat plate solar collector and wick materials and compared it to a conventional double-slope solar still. The floatable black wicks were utilized to improve the evaporation surface area and localize absorbed insolation to the evaporation surface of saline water. Results showed a 45.6% increase in productivity compared to CSS. At the system's southern and northern edges, energy efficiency has increased by 25.3% and 37.2%. According to the results, wick materials have been shown to be effective in retaining heat during periods of low sunlight. In addition to this, Morad et al., 2015 have conducted thermal research based on equations for internal and external heat transfer and energy balance to predict the performance of a double-slope solar still. In line with the thermal analysis, passive and active solar stills (solar still and solar still combined with a flat-plate solar collector) were built to use solar desalination technology to provide fresh water. The performance of both solar stills was investigated as a function of changes in the thickness of the glass cover and the depth of the brine in the basin and whether glass cover cooling (the flash method) was used or not. The findings of the experiment indicate that the utilization of active solar still with a 1 cm basin brine depth, 3 mm glass cover thickness, and flash technique cover cooling with 5 min on and 5 min off resulted in the highest freshwater productivity (10.06 l/m^2.day) and internal thermal efficiency (80.6%). In contrast, using passive solar still yielded lower freshwater productivity (7.8 l/m^2.day) and internal efficiency (57.1%).

Hassan et al., 2019 conducted an experimental study on the effect of a saline water medium on the effectiveness of a double solar still with a tracked parabolic trough collector (TPTC). Three different saline water media are considered in the basin: saline-water-soaked sand, steel wire mesh, and pure saline water. Investigations were done into how these media affected the amount of freshwater produced by both the modified solar still (solar still with TPTC) and the normal solar still. Experiments were performed in the cold winter and warm summer seasons. Using wire mesh and sand boosts the improved system's daily freshwater production by about 3.1% and 13.7%, respectively, in the winter and about 3.4% and 14.1% in the summer, according to the data. The efficiency of the redesigned system is also improved by 3.3% and 15.3% in the summer and 3.9% and 13.8% in the winter, respectively, by using wire mesh and sand with salt water.

In addition, Patel et al., 2021, summarized the water purification systems by adding several collector types to a double-slope solar still. A comparison based on yield has been conducted. Studies have been done on the many kinds of collectors augmented with DSSS. The active mode of certain additional DSSS types has also been investigated. Based on the findings of the research, it has been determined that DSSS in active mode might be a potential mitigation strategy for the current freshwater shortage problem. The yield of solar energy is still increased by integrating FPC, PTC, PV/T, solar ponds, etc.; however, the number of FPCs should not exceed 2 to 3 due to the exergy efficiency starting to decline and the yield not increasing much. Also, it has been determined that a double-slope water purification system powered by solar energy and using evacuated tubular collectors is superior to other systems of the same kind.

In further development, the evacuated tubular collector is a type of solar thermal collector with a vacuum layer between two glass tubes. The absence of a medium in a vacuum precludes convective heat transfer, thereby preventing heat loss through this mechanism. Therefore, the amount of heat introduced into the solar basin remains relatively elevated. Singh and Tiwari, 2017, presented the formulation of a characteristic equation for a double slope solar-powered water purification system that utilizes N evacuated tubular collectors. The researchers concluded that using N-evacuated tubular collectors in a double slope water purifier system based on solar energy results in higher daily energy efficiency than other comparable configurations. This is attributed to the lack of convective heat transfer.

While et al., 2018, conducted a theoretical analysis of the energy metrics of a double-slope solar distiller equipped with an N identical evacuated tubes collector. The schematic diagram of the N-ETC-DS depicts the integration of N identical evacuated tubular collectors into a double-slope solar still. The researchers found the conclusion that the life cycle conversion efficiency of a solar energy-based double slope solar distiller that integrates N-like evacuated tubular collectors is 59.42% greater than that of a solar energy-based double slope passive water purifier with an equivalent basin area. The occurrence was attributed to the thermal input into the basin by collectors. Table 1 summarizes the results of a study that compared the effectiveness of different strategies for enhancing output in double-slope solar stills.

Table 1. Summary of methods for increasing the production of a double-slope solar still

Study	Technology Used	Productivity	Remarks
Hybrid double slope SS with PCM			
Agrawal and Singh, 2022	Modified single-basin DSSS with binary eutectic PCM	0.31 kg/m^2 h	Integration of PCM enhances energy efficiency from 22.21% to 30.42%. Daily experimental efficiency is still 30.42%, while the simulated value is 33.15%. Modified SS is anticipated to produce a maximum freshwater yield of 0.31 kg/m^2.h
Suneesh et al., 2013	Built and tested a solar still with a "V" shaped basin with and without cement blocks.	24%	The system efficiency remains at 24% with cement blocks dropping to 20% without them. Heat emission from cement blocks post 5:00 PM leads to a 4% enhancement in efficiency. Cement is an effective phase change material for solar desalination. Solar still offers a viable solution for providing fresh water to rural and arid regions.
Patil and Dambal, 2016	DSSS with paraffin wax and sensible heat storage materials (black pebbles)	30%	The findings suggest that the proportion of productivity achieved for Paraffin wax and black coated tray was 30%. Conversely, the percentage productivity for black pebbles and black coated trays was 18%, and for Paraffin wax and black pebbles, it was 13%.
Murugavel et al., 2010	Energy-storing materials (quartzite rock, red brick fragments, cement concrete fragments, washed stones, and iron scraps)	2.1 kg/m^2	Energy-storing material distiller, particularly quartzite rock, boosts nighttime output. Quartzite rock of size 3/4 produced 2.1 kg/m^2 of output, an increase of 6.2% compared to no energy-storing material. Heat storage extends the working day overnight, increasing overall output. Installation and operation expenses may increase, but it is advisable for realistic applications.
Shah and Damor, 2013	Single basin DSSS with and without heat-absorbing materials (rubber and black dye)	26.7%	Output with black dye as heat-absorbing material is 26.7% higher than without absorbing materials, allowing for greater distilled water production.
Saravanan et al., 2021	Single basin DSSS with Kanchey marbles as sensible heat storage	16.32%	The yield of fresh water with Kanchey marbles is 16.32% greater than without. Double-slope solar still with marbles as heat storage produces the highest output of 4.094 l/m^2.

Study	Technology Used	Productivity	Remarks
Hybrid double slope SS with PCMs and condenser			
Faegh and Shafii, 2017	Solar still with external condenser containing PCM and heat pipes	6.56 kg/m^2	Incorporation of PCM and heat pipes increases yield by 86% compared to systems without PCM. Yield reached 6.555 kg/m^2 per day with an efficiency of 50%. The cost per liter per unit area of produced water is 0.015 USD in optimal case.
Hybrid double slope SS with PCM, collector, and fins			
Prajapati et al., 2019	Circular fins and phase change material ($Mg_2SO_4.7H_2O$)	1940 ml/m^2	The findings of the study indicate that the utilization of phase change material and aluminum fins in a solar distillation unit leads to a higher daily yield of potable water compared to traditional solar still.
Hedayati-Mehdiabadi et al., 2020)	PV/T technology and paraffin wax as a PCM	6.5 kg/m^2	Integration of PV/T technology increases basin water temperature and production during the daytime. PCM facilitates increased nocturnal production. An increase in mass flow rate leads to higher exergy efficiency. Dual collectors enhance exergy efficiency. During the summer season, the highest water production recorded was 6.5 kg/m^2.
Hybrid double slope SS with PCM and external reflector			
Gnanaraj and Velmurugan, 2022	Heat storage material in the basin and external reflector	63.97% for reflectors, 23.08 for heat storage materials	The findings indicate that the efficiency of the heat storage material distiller exceeded that of the conventional distiller by 23.08%. The implementation of a reflector in the distillation process resulted in a significant production increase of 63.97%.
Gnanaraj and Velmurugan, 2019	Heat-storage material in the basin + adding an external reflector	Various percentages higher than conventional still: 58.5%, 69.8%, 42.3%, 93.4%, and 171.4%	The estimates for production using the alternative method exceeded those of the conventional still by 58.47%, 69.84%, 42.33%, 93.39%, and 171.43%. The impact of the temperature differential between the water glass and the basin water temperature on the rate of production was examined.

Table 1. (Continued)

study	Technology Used	Productivity	Remarks
Hybrid double slope SS with various types of collector and wick materials			
Al-Molhem and Eltawil, 2020	Flat plate solar collector and wick materials	45.6%	A significant rise of 45.6% in productivity in comparison to CSS. Energy efficiency has experienced a rise of 25.3% and 37.2% at the southern and northern boundaries of the system, respectively. Utilization of wick materials has demonstrated efficacy in preserving thermal energy during instances of reduced solar radiation.
Morad et al., 2015	Flat-plate solar collector	10.06 l/m^2 per day	Active solar still with 1 cm of brine in the basin, 3 mm of glass covering it, and flash technique cover cooling for 5 minutes on and 5 minutes off produced the most freshwater (10.06 l/m^2.day) and had the highest internal thermal efficiency (80.6%). The passive solar still, in contrast, had a lower internal efficiency (57.1%) and freshwater productivity (7.8 l/m^2.day).
Hassan et al., 2019	Tracked parabolic trough collector (TPTC), saline-water-soaked sand, steel wire mesh, and pure saline water are tested in the basin.	Enhanced by 3.1%, 13.7% in winter and 3.4%, 14.1% in summer.	The utilization of wire mesh and sand in conjunction with salt water has resulted in an enhancement of the efficiency of the reconfigured system by 3.3% and 15.3% during the summer season and by 3.9% and 13.8% during the winter season, respectively.
Patel et al., 2021	Incorporating many types of collectors into a double-slope solar still	-------	Research findings have indicated that a solar-powered double-slope water purification system utilizing evacuated tubular collectors exhibits superior performance compared to other systems of a similar nature.
Singh and Tiwari, 2017	double-sloped solar still with N evacuated tubular collectors	-------	The study's findings indicate that implementing N evacuated tubular collectors in a double slope water purification system powered by solar energy yields greater daily energy efficiency than other similar setups. The absence of convective heat transfer is the reason for this phenomenon.
Singh and Al-Helal, 2018	N- identical evacuated tubular collectors	-------	The researchers concluded that the life cycle conversion efficiency of a solar-powered double-slope solar distiller with N-identical evacuated tubular collectors is 59.42% higher than that of a solar-powered double-slope passive water purifier with an equivalent basin area.

5. Summary of Benefits of Hybrid Double-Slope Solar Stills

Solar desalination using double-slope solar stills has emerged as a promising technology for producing fresh water. Hybrid double-slope solar stills offer several benefits in terms of water purification and energy efficiency. Here is a summary of the advantages associated with this type of solar still:

- Enhanced freshwater production: Hybrid double-slope solar stills are designed to maximize water production rates. By combining the features of a double-slope solar still with additional technologies or components, such as solar concentrators or energy storage substances, these systems can significantly increase the yield of purified water compared to traditional solar still designs.
- Integrating energy storage substances into solar stills is necessary to accumulate surplus solar energy and enhance the output of distilled water. These materials facilitate the utilization of stored energy during periods of low solar radiation, thereby leading to a rise in production.
- Utilizing solar fins in the solar stills leads to a larger heat transfer area on the surface, enabling better absorption of solar radiation. This reduces the time required for evaporating the saline water and enhances production rates.
- Internal modifications, such as incorporating a finned corrugated basin, heat storage material, and wick material, and external modifications, like the use of reflectors, resulted in even distribution of distillate production throughout varying sunshine hours.
- Based on previous findings, it is recommended to use TCTC with a solar still because it increases the yield of the still efficiently, is straightforward, inexpensive, and runs on free solar energy.
- Comparisons have indicated that solar-based double-slope water purification units integrated with evacuated tubular collectors outperformed other types of collectors. This highlights the superiority of evacuated tubular collectors in terms of distillate production. Nonetheless, it should be noted that the utilization of evacuated tubes in solar thermal systems may entail a higher cost in contrast to conventional flat plate collectors.
- The incorporation of PCMs in solar stills increased the average energy efficiency of modified systems to 30.42%, surpassing the

efficiency of traditional solar stills (22.2%). This indicates the significant impact of PCM integration on overall energy efficiency.

- Optimal values for the mass flow rate of saline water and the number of PV/T collectors were determined based on freshwater production and exergy analysis, ensuring desirable performance and efficiency.
- The investigation of the economic aspects of solar still outfitted with an external latent heat storage system, and evacuated tube collectors revealed that the cost per liter per square meter of the produced water is 0.015 USD under optimal conditions.

In summary, hybrid double-slope solar stills provide enhanced freshwater production, improved energy efficiency, increased water quality, versatility in application, sustainability, and cost-effectiveness. These benefits make them a promising technology for addressing water scarcity challenges and promoting sustainable water management practices.

Conclusion

The detailed analysis of literature on enhancing hybrid double-slope solar stills' production has revealed important findings. Incorporating energy storage materials, utilizing solar fins, internal and external modifications, integrating PCMs, and carefully selecting collector types and operational parameters have all been shown to enhance distillate output, performance, and thermal efficiency significantly. These conclusions provide valuable guidance for future advancements in solar desalination technologies, facilitating the development of more efficient and sustainable water production systems.

References

Agrawal, R., and Singh, K. D. P. (2022). Experimental Investigation and Computational Modeling of Double Slope Solar Still Integrated With Eutectic Phase Change Material. *Journal of Energy Storage*, 52: 104802.

Agyenim, F., Hewitt, N., Eames, P., and Smyth, M. (2010). A Review of Materials, Heat Transfer and Phase Change Problem Formulation for Latent Heat Thermal Energy Storage Systems (LHTESS). *Renewable and Sustainable Energy Reviews,* 14(2): 615-628. https://doi.org/10.1016/j.rser.2009.10.015.

Al-Kayiem, H. H. (2019). Solar Thermal: Technical Challenges And Solutions for Power Generation. *Journal of Mechanical Engineering Research and Developments*, 42(4): 269-271. http://doi.org/10.26480/jmerd.04.2019.269.271.

Al-Kayiem, H. H., Mohamed, M. M., and Syed, S. I. U. (2023). State of the Art Hybrid Solar Stills for Desalination. *Arabian Journal for Science and Engineering* 48: 5709-5755. https://doi.org/10.1007/s13369-022-07516-8.

Al-Molhem, Y. A., and Eltawil, M. A. (2020). Enhancing the Double-Slope Solar Still Performance Using Simple Solar Collector and Floatable Black Wicks. *Environmental Science and Pollution Research*, 27: 35078-35098.

Alawee, W. (2015). Improving the Productivity of Single Effect Double Slope Solar Still by Simple Modification. *Journal of Engineering*, 21(8): 50-60.

Alawee, Dhahad, H. A., Ahmed, I. S., and Mohammad, T. A. (2021). Experimental Investigation on an Elevated Basin Solar Still With Integrated Internal Reflectors And Inclined Fins. *Journal of Engineering Science and Technology*, 16(1): 762-777.

Alawee, W. H., Essa, F. A., Mohammed, S. A., Dhahad, H. A., Abdullah, A. S., Omara, Z. M., & Gamiel, Y. (2021). Improving the Performance of Pyramid Solar Distiller Using Dangled Cords of Various Wick Materials: Novel Working Mechanism of Wick. *Case Studies in Thermal Engineering*, 28: 101550. https://doi.org/10.1016/j.csite.2021.101550.

El-Sebaii, Al-Ghamdi, A. A., Al-Hazmi, F. S., and Faidah, A. S. (2009). Thermal Performance Of A Single Basin Solar Still With PCM As A Storage Medium. *Applied energy*, 86(7-8): 1187-1195. https://doi.org/10.1016/j.apenergy.2008.10.014.

Essa, F. A. (2022). Thermal *Desalination Systems: From Traditionality To Modernity And Development*. Chapter in the book: Distillation Processes-From Conventional to Reactive Distillation Modeling, Simulation and Optimization. Edited, Vilmar Steffen (edi). IntechOpen, 23. https://doi.org/10.5772/intechopen.101128.

Faegh, M., and Shafii, M. B. (2017). Experimental Investigation of a Solar Still Equipped With an External Heat Storage System Using Phase Change Materials and Heat Pipes. *Desalination*, 409: 128-135.

Fathy, M., Hassan, H., and Ahmed, M. S. (2018). Experimental Study on The Effect of Coupling Parabolic Trough Collector with Double Slope Solar Still on Its Performance. *Solar Energy*, 163: 54-61.

Gnanaraj, S. J. P., and Velmurugan, V. (2019). An Experimental Study on The Efficacy of Modifications in Enhancing The Performance of Single Basin Double Slope Solar Still. *Desalination*, 467: 12-28.

Gnanaraj, S. J. P., and Velmurugan, V. (2022). Experimental Investigation on The Performance of Modified Single Basin Double Slope Solar Stills. *International Journal of Ambient Energy*, 43(1): 206-215.

Gupta, B., Kumar, A., and Baredar, P. V. (2017). Experimental Investigation on Modified Solar Still Using Nanoparticles and Water Sprinkler Attachment. *Frontiers in Materials*, 4(23).

Hammoodi, K. A., Dhahad, H. A., Alawee, W. H., and Omara, Z. M. (2023). A Detailed Review of the Factors Impacting Pyramid Type Solar Still Performance. *Alexandria Engineering Journal*, 66: 123-154. https://doi.org/10.1016/j.aej.2022.12.006.

Hassan, H., Ahmed, M. S., and Fathy, M. (2019). Experimental Work on the Effect of Saline Water Medium on the Performance Of Solar Still With Tracked Parabolic Trough Collector (TPTC). *Renewable Energy,* 135: 136-147. https://doi.org/10.1016/j.renene.2018.11.11.

Hedayati-Mehdiabadi, E., Sarhaddi, F., and Sobhnamayan, F. (2020). Exergy Performance Evaluation of A Basin-Type Double-Slope Solar Still Equipped With Phase-Change Material And PV/T Collector. *Renewable Energy*, 145: 2409-2425.

Janarthanan, B., Chandrasekaran, J., and Kumar, S. (2006). Performance of Floating Cum Tilted-Wick Type Solar Still With The Effect of Water Flowing over the Glass Cover. *Desalination*, 190(1-3): 51-62.

Kabeel, A., Arunkumar, T., Denkenberger, D., & Sathyamurthy, R. (2017). Performance enhancement of solar still through efficient heat exchange mechanism–a review. *Applied Thermal Engineering*, 114: 815-836.

Morad, M., El-Maghawry, H. A., and Wasfy, K. I. (2015). Improving the Double Slope Solar Still Performance by Using Flat-Plate Solar Collector and Cooling Glass Cover. *Desalination*, 373: 1-9.

Murugavel, Sivakumar, S., Ahamed, J. R., Chockalingam, Kn., and Srithar, K. (2010). Single Basin Double Slope Solar Still With Minimum Basin Depth and Energy Storing Materials. *Applied energy*, 87(2): 514-523. https://doi.org/10.1016/j.apenergy.2009.07.023.

Nayi, K. H., and Modi, K. V. (2018). Pyramid Solar Still: A Comprehensive Review. *Renewable and Sustainable Energy Reviews*, 81:136-148.

Pandey, G. C. (1983). Effect of Dye on The Performance of A Double Basin Solar Still. *International journal of energy research*, 7(4): 327-332.

Patel, R. V., Bharti, K., Singh, G., Mittal, G., Singh, D. B., Yadav, A. (2021). Comparative Investigation of Double Slope Solar Still by Incorporating Different Types of Collectors: A Mini Review. *Materials Today: Proceedings,* 38: 300-304. https://doi.org/10.1016/j.matpr.2020.07.338.

Patil, B. K., and Dambal, S. (2016). Design and Experimental Performance Analysis of Solar Still Using Phase Changing Materials and Sensible Heat Elements. I*nternational Journal of Research in Mechanical Engineering & Technology* 6: 144-149.

Prajapati, U. K., Gupta, B., and Bhalavi, J. (2019). Performance Improvement of Double Slope Solar Still Using Aluminum Fins and Phase Change Material. *Journal of emerging technologies and innovative research*, 6: 295-302.

Rajaseenivasan, T., Kalidasa Murugavel, K., and Elango, T. (2015). Performance and Exergy Analysis of A Double-Basin Solar Still with Different Materials In Basin. *Desalination and Water Treatment*, 55(7): 1786-1794.

Saravanan, N.M., Rajakumar, S., and Moshi, A. A. M. (2021). Experimental Investigation on The Performance Enhancement of Single Basin Double Slope Solar Still Using Kanchey Marbles as Sensible Heat Storage Materials. *Materials Today: Proceedings*, 39: 1600-1604.

Selvaraj, K., and Natarajan, A. (2018). Factors Influencing the Performance and Productivity of Solar Stills-A Review. *Desalination* 435, 181-187. https://doi.org/10.1016/j.desal.2017.09.031.

Shah, R., and Damor, A. (2013). Performance Improvement of Double Slope Solar Still Using Heat Absorbing Materials. *International Journal of Mechanical and Production Engineering Research and Development*, 3: 143-148. https://doi.org/10.1016/j.enbuild.2020.110141.

Sharshir, S. W., Elkadeem, M., and Meng, A. (2020). Performance Enhancement of Pyramid Solar Distiller Using Nanofluid Integrated With V-Corrugated Absorber And Wick: An Experimental Study. *Applied Thermal Engineering*, 168: 114848. https://doi.org/10.1016/j.applthermaleng.2019.114848.

Singh, D., and Al-Helal, I. (2018). Energy Metrics Analysis of N Identical Evacuated Tubular Collectors Integrated Double Slope Solar Still. *Desalination*, 432: 10-22.

Singh, D. B., and Tiwari, G. N. (2017). Analytical Characteristic Equation of N Identical Evacuated Tubular Collectors Integrated Double Slope Solar Still. *Journal of Solar Energy Engineering*, 139(5). https://doi.org/10.1115/1.4036855.

Suneesh, P., Jayaprakash, R., and Sanjay, K. (2013). Solar Desalination-Effect of Cement Absorber in Double Slope Solar Still. *International Journal of Applied Engineering Research and Development*, 3(4): 11-18.

Tuly, S., Rahman, M., Sarker, M., & Beg, R. (2021). Combined Influence of Fin, Phase Change Material, Wick, And External Condenser on The Thermal Performance of a Double Slope Solar Still. *Journal of Cleaner Production*, 287: 125458. https://doi.org/10.1016/j.jclepro.2020.125458.

Tuly, S., Sarker, M., Das, B. K., and Rahman, M. S. (2021). Effects of Design And Operational Parameters on The Performance of A Solar Distillation System: A Comprehensive Review. *Groundwater for Sustainable Development*, 14: 100599. https://doi.org/10.1016/j.gsd.2021.100599.

Chapter 9

Solar-Powered Atmospheric Water Generation Using Photovoltaic Panels

Ammar M. Al-Tajer[1,*]
Wissam H. Alawee[2]
Hayder A. Dhahad[1]
and Zakaria. M. Omara[3]

[1]Mechanical Engineering Department, University of Technology- Iraq, Baghdad, Iraq
[2]Control and Systems Engineering Department, University of Technology- Iraq, Baghdad, Iraq
Training and Workshops Center, University of Technology- Iraq, Baghdad, Iraq
[3]Mechanical Engineering Department, Faculty of Engineering, Kafrelsheikh University, Kafrelsheikh, Egypt

Abstract

This chapter explores solar energy and thermoelectric technology for sustainable water production, especially in remote and off-grid areas with limited access to electricity. Solar energy offers a cost-effective and eco-friendly solution, eliminating the need for grid connection as technology costs decrease. Solar thermal desalination systems are particularly advantageous, generating valuable heat during the desalination process, which can be used for space or process heating or converted into electricity through thermoelectric generators. Integration of thermoelectric generators into atmospheric water generation (AWG) systems enhances water production capabilities, even in regions with low humidity or high temperatures, such as arid or desert areas. AWG systems equipped with thermoelectric generators can be tailored for

* Corresponding Author's Email: me.21.16@grad.uotechnology.edu.iq.

In: Advancements in Sustainability Systems
Editors: Hussain H. Al-Kayiem, Hasanain A. Abdul Wahhab and Raed A, Jessam
ISBN: 979-8-89113-571-0

various capacities, from small units serving daily water needs to large-scale systems for entire communities, effectively addressing water scarcity challenges and providing clean drinking water. The paper highlights three highly efficient water production technologies based on thermoelectric principles: Humidification-Dehumidification (HDH) method, optimized Solar-Driven Thermoelectric Dehumidification system, and Thermoelectric-Photovoltaic system of humid air, each with specific production costs per liter.

Keywords: atmospheric water generation (AWG), vapor concentration. sorbent sorption (VCSS), thermoelectric

Nomenclature

PV	Photovoltaic
HDH	Humidification-Dehumidification
AWG	Atmospheric Water Generation
VCSS	The Vapor Concentration Sorbent Sorption
SWSs	Selective Water Sorbents
AC	Alternating Current
EES	Engineering Equation Solver
COP	coefficient of performance
DC	Direct current
CFD	Computational Fluid Dynamics
WHO	World Health Organization
TEC	Thermoelectric cooler
TFWG	Thermoelectric Freshwater generator
PVT	Photovoltaic Thermal.

1. Introduction

The scarcity of freshwater has become a pressing global issue, impacting human survival and economic development. Factors such as increasing demand, pollution, and mismanagement have contributed to the freshwater crisis (Smith, 2003). Pollution from industrialization, urbanization, and agriculture poses a significant threat to freshwater sources by introducing toxic chemicals, fertilizers, and pesticides that contaminate rivers, lakes, and groundwater, rendering them unsafe for consumption. Furthermore, plastic pollution has emerged as a modern challenge, with microplastics infiltrating freshwater systems, potentially harming aquatic life and posing risks to human health. Climate change exacerbates the freshwater crisis by altering rainfall patterns, increasing temperatures, and intensifying natural disasters like droughts and floods, affecting both freshwater availability and quality (Chaudhry and Malik, 2017; Sandford and Adeel, 2012; Jury and Vaux, 2007).

Al-Karaghouli and Kazmerski, 2013 investigated two primary technologies employed in water desalination are explained: thermal (phase-change) processes and membrane processes. The paragraph highlights that energy costs are crucial in determining these processes' overall water production cost, constituting approximately 50% of the total cost. To address energy-related concerns, the discussion revolves around the potential of renewable energy systems such as solar thermal, solar photovoltaic, wind, and geothermal technologies, which offer promising solutions for desalination applications, especially in remote regions facing water scarcity.

Sun et al., 2023 investigated the potential of artificial water generation processes harnessing readily available solar energy to combat global energy and water shortages. It emphasizes the significance of aerogels in water generation from brine and atmospheric water due to their unique characteristics, including photo-thermal conversion and water transport capabilities. Also, they reviewed various structural designs to enhance evaporation performance in solar-driven water production systems. However, challenges associated with the commercialization and widespread adoption of aerogel-based water production systems are also discussed, including their disconnection from the current aerogel industry, high production costs, weak mechanical properties, and the lack of standardized performance testing. Despite these challenges, they concluded a positive outlook on the potential application opportunities of aerogel-based systems.

Addressing the freshwater crisis requires implementing effective water management policies, pollution reduction strategies, and promoting renewable energy sources. Using renewable energy sources, such as those illustrated in Table 1, can employ various water production methods (Salehi et al., 2020; Wang et al., 2018; Khalil et al., 2016).

Renewable energy technologies have the potential to make a substantial contribution to generating clean and safe water for communities worldwide (El-Ghonemy, 2012; Sleiti et al., 2021; Esfahani et al., 2011; Kabeel et al., 2015). Solar energy is a renewable and environmentally friendly energy source with diverse applications. One prominent use of solar energy involves the generation of electricity using solar panels, which convert solar radiation into usable electrical power for homes, businesses, and various facilities.

Table 1. Renewable energies for water production techniques

Technique	Description
Desalination	This process involves removing salt and other minerals from seawater to make it drinkable.
Solar stills	Devices use the sun's heat to evaporate water, which is then condensed and collected as fresh water.
Rainwater harvesting	This technique involves collecting and storing rainwater for later use.
Fog harvesting	This method involves collecting water from fog or dew using mesh or netting.
Geothermal water production	Geothermal water production

Salehi et al., 2020 focus on technologies developed to produce drinking water from humid air, but the main concern lies in the high energy consumption associated with these systems. To tackle this challenge, they proposed the adoption of renewable energy sources, particularly solar and wind energy, as a solution to reduce energy costs in the water extraction process from humid air.

Solar energy can be used to produce water from the air using humidification-dehumidification. This process involves using solar energy to heat the air, causing it to rise and subsequently cool (Jabari et al., 2020; Dave et al., 2021; Almusaied and Asiabanpour, 2017). As the air cools, the water vapor within it condenses, allowing for clean water collection. This method proves beneficial in regions with low humidity or limited access to traditional water sources. Overall, solar energy stands as a versatile and sustainable solution, capable of meeting diverse energy demands while also contributing to clean water production (Hosseini et al., 2013; Mekhilef et al., 2011). Various processes utilizing solar energy for water production from atmospheric air are depicted in Figure 1.

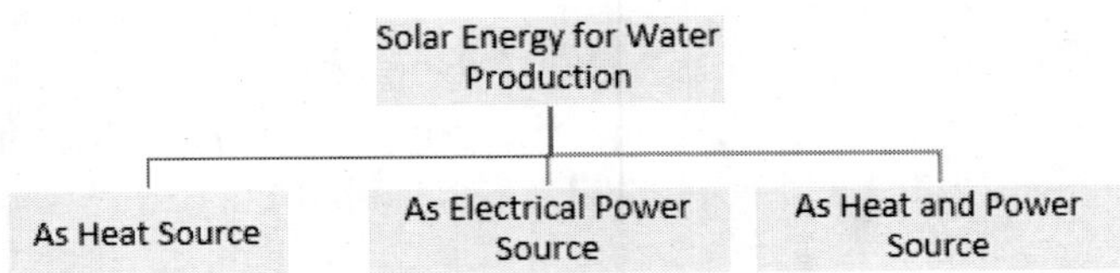

Figure 1. Solar energy conversion methods for atmospheric water production.

Atmospheric water generation (AWG) involves extracting water from the air using energy, and one effective method for achieving this is by harnessing solar energy. Various devices can be employed to carry out this process. The present study will specifically investigate the techniques utilized for water generation powered by photovoltaic (PV) panels, as illustrated in Figure 1.

Using solar energy to extract water from the atmosphere represents an innovative and sustainable solution that effectively tackles several pressing challenges of our time, including water scarcity and environmental sustainability. This cutting-edge technology, referred to AWG offers a multitude of advantages:

1. Water Scarcity Mitigation: As global water scarcity becomes increasingly urgent, AWG systems provide a localized and decentralized approach to water production.
2. Reduced Energy Costs: The energy expenses associated with traditional water sources like groundwater pumping or desalination are markedly diminished by employing solar energy for AWG.
3. Low Environmental Impact: In comparison to certain conventional water extraction methods, AWG systems have a relatively modest environmental footprint.
4. Resilience to Climate Change: AWG systems exhibit resilience in the face of more frequent and severe droughts and water shortages resulting from climate change. They depend on humidity levels in the air instead of precipitation or groundwater availability.
5. Community Empowerment: Solar-powered AWG systems are adaptable and can be deployed at various scales, ranging from individual households to larger community installations.
6. Reduced Pressure on Existing Water Sources: By offering an additional source of clean water.

2. Solar-Powered Atmospheric Water Generator

A feasible solution is to utilize solar energy for generating electricity, specifically for operating a device that extracts water from the air. This innovative technology is called a solar-powered AWG. It converts solar radiation into electricity through solar panels, subsequently used to power a system that cools the air and transforms water vapor into liquid. This process can produce clean and safe drinking water in regions facing water scarcity

(Kode et al., 2022; Micari et al., 2023). This chapter identifies the technologies that rely on solar energy as their power source for extracting water from the air, as shown in Figure 2.

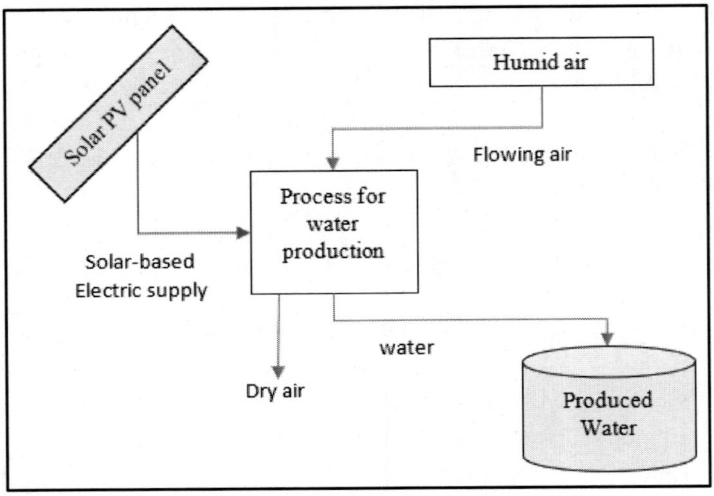

Figure 2. Fundamental of solar-assisted water production from ambient air.

2.1. Refrigeration System

In AWG systems, refrigeration systems play a vital role by supplying the necessary cooling to condense water vapor from the air. These systems have the flexibility to be powered by various energy sources, including electricity powered by solar energy (PV). Table 2 compares the different techniques used for refrigeration for water production.

Cabacungan et al., 2009 conducted an experimental study to explore the potential of solar energy in AWG and rainwater purification. Their research aimed to develop an affordable AWG system. Their system involved utilizing a car air-conditioning unit as a dehumidifier, with the car compressors replaced by a 1.5 horsepower compressor. To power the system, they employed a PV panel, which was connected to inverters and batteries, generating AC 220 V and 10.8 A. Moist air was directed to the evaporator for water production, as illustrated in the block diagram in Figure 3. The system achieved a water production rate of 18.3 L/20 hr. with an average relative humidity of 69.2%. The power consumption was measured at 2 kilowatt-hours per liter of water produced.

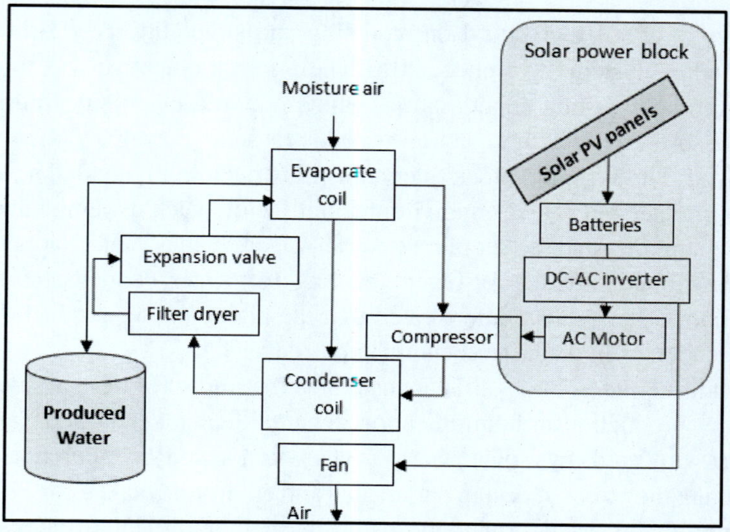

Figure 3. Block diagram outlining Cabacungan et al. 2009 method for AWG using dehumidifier device.

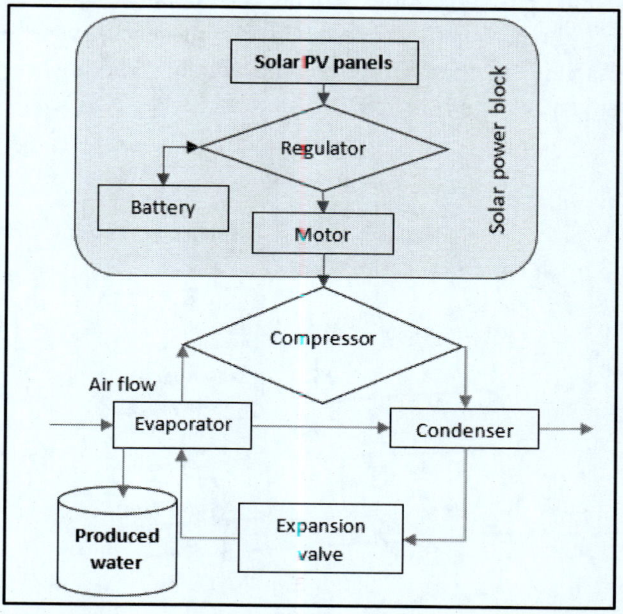

Figure 4. Block diagram outlining Aye et al. 2013 method using PV-driven system for AWG.

Aye et al., 2013 focused on modeling and simulating AWG through a solar chilled system. Aye et al., 2013 The primary objective was to extract water through cooling condensation, which required electricity to drive the compressor. To meet the electricity needs, a solar power PV system was employed. The system's block diagram is illustrated in Figure 4. The amount of water generated relied on environmental factors such as temperature and relative humidity and the supplied electricity from the PV panel to operate the compressor and achieve sufficient cooling for water condensation in the evaporator. The results demonstrated that the highest water production from the air occurred in August, reaching 36.6 L/day.

Yildirim and Solmuş, 2014 investigated the impact of design parameters on water production in humidification dehumidification (HDH) desalination systems powered by solar energy. They conducted a theoretical study examining the effect of design parameters and environmental factors, depicted in Figure 5, on water production using the dehumidification theory in the atmospheric conditions of Antalya, Turkey. The study revealed that the heating of freshwater directly influences the quality of clean water, whereas heating the air has negligible effects on potable water productivity. Additionally, a higher air flow rate positively influences water productivity, but there is a limit to the airflow beyond which it diminishes water yield (Mulchandani and Westerhoff, 2020).

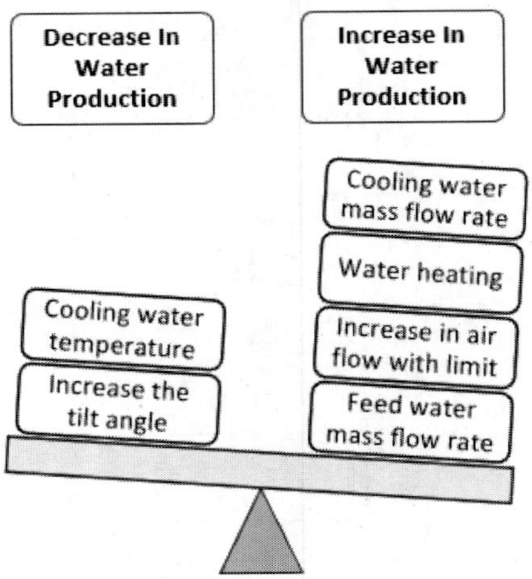

Figure 5. Identified design parameters influencing freshwater production.

Elattar et al., 2016 investigated a novel hybrid air conditioning system that generates fresh water using the humidification–dehumidification process powered by solar energy. They conducted a theoretical study, employing a mathematical model based on thermodynamic analysis and utilizing EES software and C++ programming to examine the impact of various design parameters on cooling and freshwater productivity.

As depicted in Figure 6, the system of Elattar et al., 2016 consisted of a solar air collector that heated the air to a specific temperature. The heated air then passed through a heat exchanger to reach the desired temperature and proceeded to the humidification device. The hot air was mixed with previously heated water from a solar water collector in the humidification device, increasing the air's specific humidity. The air-water mixture was then directed to the dehumidification section, where the moist air was cooled to the point of dew drop formation, resulting in the collection of fresh water. The resulting air and recirculated air were used in the cooling process. The freshwater was chemically treated and stored as potable water. The study's findings indicated that the water productivity, cooling capacity, and power consumption of the system increase when the ambient temperature, relative humidity, and temperature difference between the system and surroundings are elevated. Moreover, the system's coefficient of performance (COP) improves with higher ambient relative humidity, lower ambient temperature, increased air flow rate, and larger temperature differences across the heating system.

Alahmer et al., 2018 focused on utilizing solar energy to generate fresh water from the air through an experimental approach. The investigation examined the water production process, which relied on intensifying the water vapor in the air. The device involved several stages:

- An electrostatic filter was employed to remove particulate matter from the air.
- The moist air then passed through a condenser, which utilized air cooling to reduce the air's ability to hold water vapor, resulting in condensation.
- Ultraviolet rays were used to kill germs in the resulting water, lasting 30 minutes.
- The mixture underwent a sediment sieve to eliminate chemical suspensions present in the liquid.
- Water was filtered through a filtering membrane.

- The obtained product was stored in reservoirs, and the process was repeated until a sufficient quantity suitable for human use was obtained.

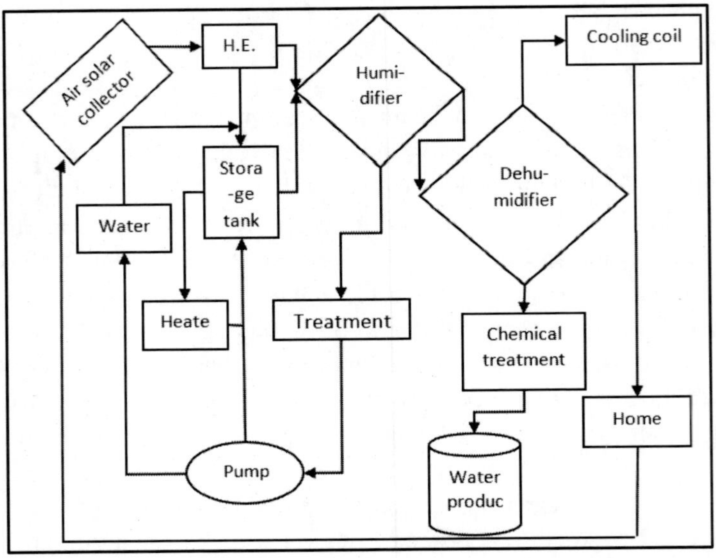

Figure 6. Block diagram outlining the proposed system by Elattar et al., 2016 for solar-driven AWG.

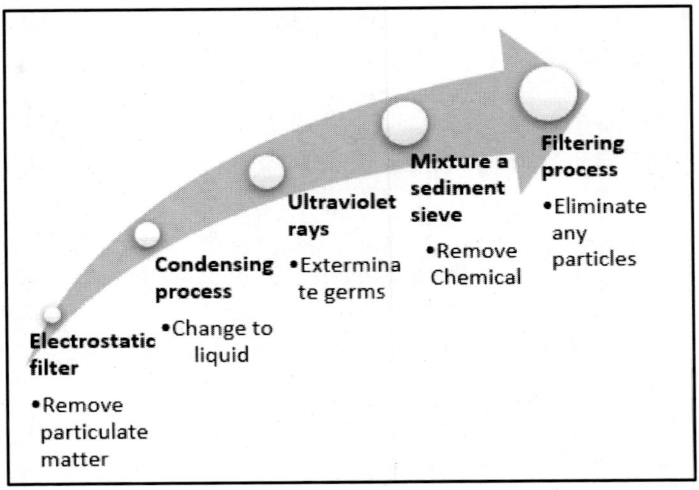

Figure 7. Steps of the water purification process.

Table 2. Comparison between refrigeration techniques for water generation

Research Study	Methodology and System Details	Key Findings
Cabacungan et al., 2009	A car air-conditioning unit is utilized as a dehumidifier.	Water production rate of 18.3 L/20 hr with 69.2% average RH.
	Car compressors are replaced with a 1.5 horsepower compressor.	Power consumption is measured at 2 kWh per liter of water produced.
	A PV panel generating AC 220 V and 10.8 A is used to power the system.	Developed an affordable AWG system.
Aye et al., 2013	Employed a solar chilled system for cooling condensation.	Highest water production occurred in August, reaching 36.6 L/day.
	Relied on environmental factors such as temperature and relative humidity.	Water generation is influenced by PV panel-supplied electricity for compressor operation.
	Used a solar power PV system to meet electricity needs.	
Yildirim et al., 2014	Investigated humidification dehumidification (HDH) desalination systems.	Heating fresh water has a direct impact on clean water quality.
	Analyzed design parameters and environmental factors.	Heating air had negligible effects on potable water productivity.
	Examined impact on water production in Antalya, Turkey's atmospheric conditions.	Higher air flow rates positively influenced water productivity, but there is a limit.
Elattar et al., 2016	Developed a novel hybrid air conditioning system using the humidification–dehumidification process.	Ambient temperature, relative humidity, and temperature difference influenced system performance.
	Employed mathematical model and thermodynamic analysis.	Higher ambient relative humidity, lower ambient temperature, and increased airflow rate improved system performance.
	Utilized EES software and C++ programming for analysis.	Achieved improved coefficient of performance (COP) with optimized conditions.
Alahmer et. Al., 2018	Experimental approach with various stages for water production.	AWG system produced high-quality potable water.
	Used electrostatic filter, condenser, UV disinfection, and filtering membrane.	Achieved a remarkable daily water production of 1722 liters using solar energy.
Continue Alahmer et. Al., 2018	Solar energy-powered refrigeration machine for water droplet formation.	Nanomaterials in heat exchangers enabled effective extraction at low humidity and temperatures.
	Examined the influence of temperature, relative humidity, air volume, and refrigeration capacity on water yield.	

Several factors influenced the water yield, including temperature, relative humidity, air volume, and the capacity of the refrigeration machine. The refrigeration machine, powered by environmentally friendly solar energy, extracted sufficient heat to form water droplets. The process of water purification is shown in Figure 7. The AWG system achieved a remarkable daily production rate of 1722 liters while consuming 200 kWh of power.

Alahmer et al., 2018 analyzed field tests on an atmospheric water extractor and found that ambient air temperature had the most significant influence on water extraction. The use of nanomaterials in the heat exchanger enabled effective extraction at low humidity and temperatures.

2.2. Thermoelectrical

Thermoelectric devices are compact, noiseless, and operate using DC electrical power. They are lightweight and simple devices based on the Seebeck and Peltier effects. These effects allow for both cooling and heating sides, which can be utilized for water production from the air. Thermoelectric devices enable the condensation of water vapor by cooling humid air below the dew point temperature (Elgendi et al., 2023; Panchal., et al., 2020; Sun et al., 2023).

Several researchers have conducted studies to showcase the potential and utilization of thermoelectric devices in extracting water from the air using solar energy. They have investigated various aspects such as process, theory, and numerical effects, considering technologies, cost of water production, and recent advancements. These studies have been compiled and summarized in Table (4). The key parameters influencing water production in thermoelectric systems, as illustrated in Table 3 (Liu et al., 2021; Wehbi., et al., 2023).

Table 3. Parameters affecting the water production from the air

Parameter	Sub-parameter
Environment Property "Climate Conditions"	• Relative humidity • Environment temperature • Altitude
Types of Thermoelectrical	• Thermal conductivity • Thermoelectric material
Electrical Supply	• Amount of Electrical Suppled in Watt • Electrical conductivity
System design	• Efficiency • Geometries • Materials

Vián et al., 2002 Researchers conducted a numerical modeling study on a thermoelectric dehumidifier that was designed and built based on thermoelectric cooling technology. The study focused on investigating the thermoelectric effect of heat transfer and phase change of the material during the condensation process on solid surfaces with temperatures below the dew point. The experimental prototype device was used to continuously create water by supplying electricity ranging from 0.0 to 100.0 W. In their research, the authors aimed to assess the feasibility of applying thermoelectric technology to a low-power dehumidifying device. The results indicated that with a 12V DC power supply and a fan airflow rate of 0.081 m^3/s, the device produced 1.4 liters of water per day.

Esfe et al., 2021 focused on freshwater production from atmospheric air. Researchers investigated the impact of the Peltier effect numerically. Their study aimed to optimize the condensation process and enhance water productivity by altering the surface temperature of thermoelectric devices. The wall temperature varied between 1 to 12 degrees Celsius, and relative humidity ranged from 35% to 80%. The findings revealed a direct correlation between decreased thermoelectric wall temperature and increased water production and condensation. The maximum water production was achieved at the lowest temperature of the cold surface, with a rate of 0.323 kg/m^2.hr. Notably, a temperature reduction of 11 degrees Celsius on the cold side of the thermoelectric cooling system led to a remarkable 36.8% increase in the water production rate.

In their experimental study, Milani et al., 2011 investigated the performance of thermoelectric devices in a dehumidification system for water production from the air. The primary objective of the research was to assess the feasibility of utilizing thermoelectric devices in the dehumidification process to cool ambient air and extract fresh water from its moisture. Psychrometric analysis was employed to calculate the water production in the study. By evaluating the performance and viability of thermoelectric devices in this context, the researchers aimed to contribute to developing efficient and sustainable water generation technologies.

Jradi et al., 2011 developed an integrated thermoelectric, photovoltaic system designed for AWG. The objective of his study was to create a combined system that incorporates thermoelectric and solar distillation technologies for dehumidification powered by solar energy. The research focused on a low-humidity region in Lebanon, and a robust model was employed to simulate the dehumidification process. As depicted in Figure 8, the testing rig featured a cooling channel where thermoelectric devices were operated using solar energy to facilitate the cooling process and generate fresh water. Additionally, a solar still produced water from the humid air with high moisture content. The system achieved a minimum water production of 10 liters per day, with the maximum amount of fresh water being generated during the system's operation between 3:00 and 6:00 p.m. for five months. Theoretical analysis indicated that the estimated cost of the generated water starts from 82 $/kL, depending on climate conditions. The study also highlighted the environmental impact of water production from the air.

Jradi and Ghali, 2012 focused on developing an optimized Solar-Driven Thermoelectric Dehumidification system. Their work involved analyzing the parameters that affect freshwater production using a combination of solar distillation and a solar-driven thermoelectric device. They examined how ambient conditions and electric current input influence the system's water-generated performance. The findings reveal that the system could produce 10 liters of fresh water daily. The optimal operation of the system consumes a total energy ranging from 5.32 kWh/day in October to 6.73 kWh/day in June. Although the initial investment cost of the system is relatively high, it offers long-term cost savings.

A prototype device was developed for harvesting water to water young trees by Muñoz-García et al., 2013. They utilized Peltier modules powered by photovoltaic solar energy. This project aimed to address the challenges of arid climates and ensure the growth and preservation of young plants. The device was designed to capture water molecules and use solar energy as a sustainable power source, prioritizing environmental friendliness. The experiment aimed to evaluate the effectiveness of using thermoelectric devices for water production and determine the optimal operating conditions. The study identified a correlation between the duration of energy supply and the water requirements, particularly during periods of drought or severe weather. When a 30-watt power source, combined with a 17Ah battery and 12 volts, was used to operate the thermoelectric device, 10 mL of water was obtained.

A prototype of a thermoelectric cooler system was developed by Suryaningsih and Nurhilal in 2016 to generate atmospheric water using renewable solar-driven power. The system, depicted in Figure 9, was optimized using CFD simulation and analysis to improve the design in the flow region. The CFD model solved the Navier-Stokes equations, including continuity, momentum, and energy equations. The research's primary focus was to ensure the system was suitable for solar energy-driven power and meet the minimum water production requirement set by the World Health Organization (WHO) at 2.5 liters per capita per day. The results demonstrated that at a temperature of 26°C, relative humidity of 69%, and an input power of 125.5 W, the system successfully produced 14.6 liters of fresh water per day from atmospheric moisture.

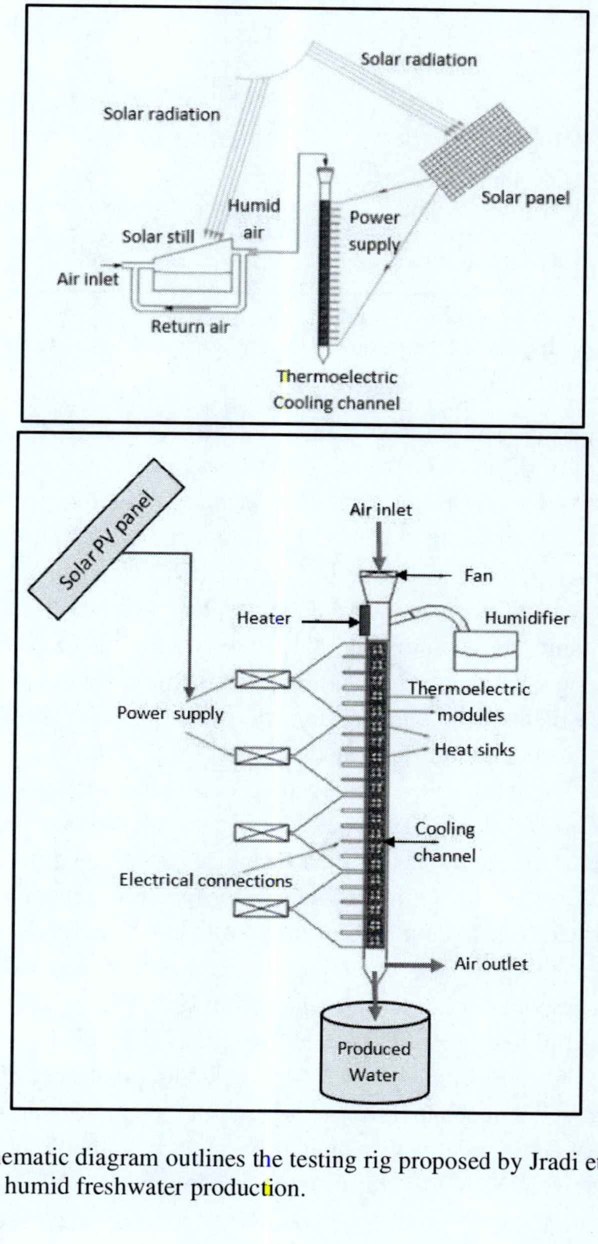

Figure 8. Schematic diagram outlines the testing rig proposed by Jradi et al., 2011 for water-low humid freshwater production.

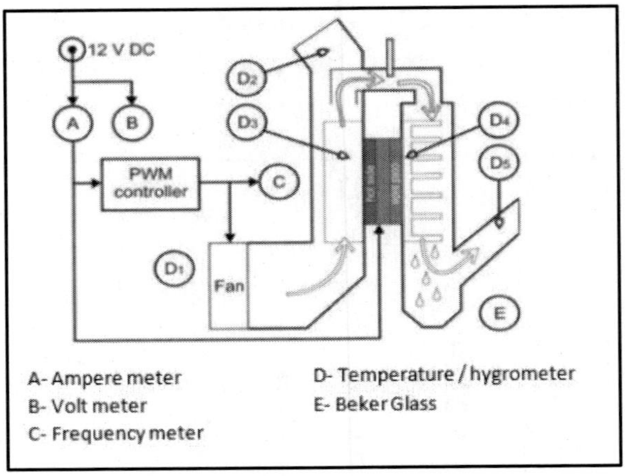

Figure 9. Block diagram of the solar-driven prototype developed by Suryaningsih and Nurhilal, 2016 of AWG.

Using thermoelectric technology, Kabeel et al., 2016 conducted a numerical study of a straightforward design approach for recovering water from the surrounding environment. The primary objective of their research was to develop a simple and efficient solution for harvesting fresh water in water-scarce regions. The prototype system for water generation relies on solar energy as its primary power source. The numerical simulation was conducted using the commercial CFD software Star-CCM+. The study examined several parameters, including pressure drop, water productivity, and the influence of ambient temperature and humidity. The simulation results showed that the dehumidification duct was 3.9, 2.679, and 2.698 L/h/m², respectively. These rates were achieved at solar power inputs of 75.61, 98.53, and 95.2 W, respectively. The mass flow rates for cases 1, 2, and 3 were calculated as 2.153×10^{-4}, 2.189×10^{-4}, and 2.226×10^{-4} kg/s, respectively.

Joshi et al., 2017 investigated freshwater generation using a thermoelectric device that operated with a cooling surface below the dew point temperature. The primary objective of their study was to develop an experimental device for harvesting water from atmospheric air, specifically in regions with humidity levels exceeding 60%. Figure 10 illustrates the experimental testing setup, featuring 10 vertically positioned thermoelectric devices arranged in a channel. At the top end of the channel, an axial fan with a flow rate of 0.014 m³/s was installed. Each set of 4 thermoelectric devices was powered by a DC power source rated at 30 V and 2 A. The results showed that the maximum water production was 240 ml for 10 hours and the maximum

COP of 0.437, corresponding to an air mass flow rate of 0.011 kg/s and relative humidity of 90%.

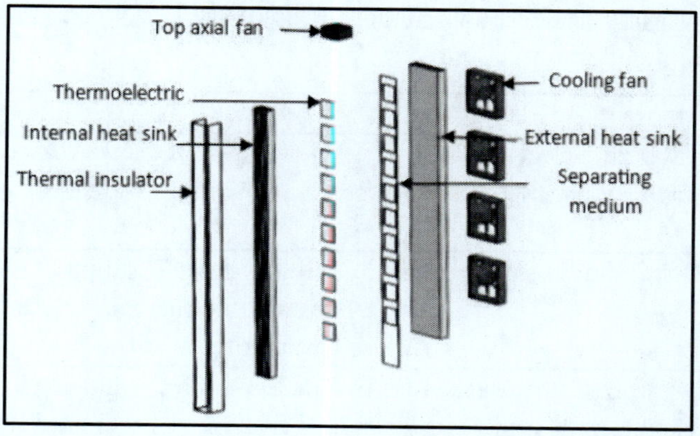

Figure 10. Schematic diagram of the experimental testing rig developed by Joshi et al., 2017 using thermoelectric AWG.

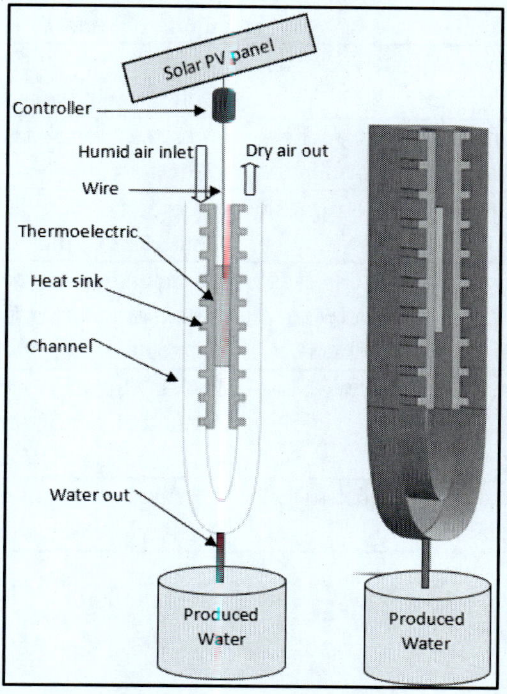

Figure 11. Schematic diagram outlines the method of Atta, 2017 using a solar and thermoelectric integrated system for AWG.

Table 4. Summary of research using a thermoelectrical device for water production.

References	Type of Study	Technique	Research focus	Capacity
Vián, et al., 2002	Mathematical modeling	Calculation model based on electric analogy, Thermoelectric dehumidifier consists of two stages	Various Peltier pellet, air flow, and voltages to study COP and condensed water capacity.	Numerical: 0.969 l/day; Experimenta 1.4 l/day
Esfe et al., 2021	Theoretical	TEC devices and dehumidification systems	Feasibility of using TEC devices in dehumidification to refrigerate ambient air and provide freshwater	0.323 kg/m^2.hr
Milani et al., 2014	Experimental and Theoretical	low-power dehumidifying device driven by solar PV panels to generate water in the Lebanese humid climate.	Parametric analysis of the performance of a solar-driven thermoelectric system to dehumidify air and produce fresh water	10 l/ day
Iradi, 2011	Experimental and Numerical	combined solar distiller and humidifying ambient air to enhance distillate for a residential application.	Parametric analysis of the performance of a solar-driven thermoelectric system to dehumidify air and produce fresh water	10 l/ day
Iradi and Ghali, 2012	Experimental	Nonconventional method using young trees and Peltier modules powered by PV.	water production by dry-climate woody crops to overcome their critical growing stage	10 ml/hour
Muñoz-García et al., 2013	Experimental and Theoretical	Thermoelectric units of Peltier module. CFD is utilized to optimize the process in the flow region.	presents the method to develop a prototype of an AWG based on a Thermo-electric cooler (TEC)	10 mL per use
Suryaningsih et al., 2016	Experimental	CFD is utilized to optimize the design process in the flow region	Development of an AWG prototype based on Thermo-electric cooler (TEC)	14.6 l/day
Kabeel et al., 2016	A numerical study	3D simulation using commercial CFD software called Star-CCM+, two-phase flow field and condensation inside the dehumidification	find a simple and effective solution of water production in remote areas where pure water using a thermoelectric device	3.9 l/m^2.h
Joshi et al., 2017	Experimental	Cooling channel along with ten thermoelectric modules	Fresh water generator based on the fundamental of Thermoelectric Cooling Effect by condensing the moisture from the ambient moist air	240 ml/10 hours
Atta, 2017	Experimental	Solar PV/batter Thermoelectric dehumidifying systems are the PV cell	A solar water condensation system built using a TE cooler, solar panels, heat exchange unit	1.0 l/h during the daylight

Atta, 2017 presented an experimental investigation into using solar energy and a thermoelectric cooling device for water production in humid areas, particularly close to the sea. The experimental setup, illustrated in Figure 11, comprised a heat exchange unit and an air circulation unit. The thermoelectric device's operation was facilitated by a 12V PV panel with a maximum output power of 120 W. The PV panel powered a DC 12V fan that facilitated the flow of moist air at a rate of 500 cubic feet per minute. The moist air was first directed to the thermoelectric device's hot side to raise the air temperature while cooling it. Subsequently, the atmospheric air was guided to the cold side of the device, where the condensation of water vapor occurred. The results demonstrated that the system produced 1.0 liters of condensed water per hour during daylight hours.

The chapter summarizes the so far reported attempts for AWG using thermoelectrical units. The compilation and main features of each system are shown in Table 4.

2.3. Solar Energy as an Electrical and Heat Source for AWG

Utilizing solar radiation as an energy and heat source for water production can be a promising approach to assess the suitability of these systems for specific regions or circumstances.

In their study, Milani et al., 2014 focused on modeling a dehumidification system that employed solar-assisted desiccant for water generation. The research had two main objectives: utilizing solar energy to produce water from the air and reducing latent heat through pre-treatment of the air feed stream for air conditioning purposes. Illustrates the system, which involved a desiccant wheel using silica gel to dehumidify the air with the assistance of PV solar energy. The atmospheric air enters the desiccant dehumidifier wheel in a specific state and then proceeds to the heat cooling exchanger, where water production occurs. The water-absorbent material releases heat into the surrounding air, producing drier air. The desiccant wheel's role is to saturate the air with water vapor, facilitating its extraction through heat exchange. The results indicate that even a small amount of silica gel can generate a water production rate of 5.2 liters per day. Furthermore, the annual water generation for London, Sydney, and Abu Dhabi was estimated as 10 kL, 13.8 kL, and 18.5 kL, respectively.

3. Economic Comparison of AWG Technologies

The specific cost of water production from a particular technology will depend on several factors, including the efficiency of the technology, the cost of energy, and the cost of the materials used to construct the technology. The cost of water production from different technologies can vary greatly, as shown in Table 5. Prediction of the cost per liter counts for the utilized area, the equipment, the workforce for maintenance and running, and the replacement of degraded parts. It is important to note that the cost of water production from these technologies will likely decrease as the technology matures and the cost of the materials used to construct the system decreases. This suggests that these technologies could become more cost-effective options for water production in the future.

Table 5. Water production cost

Author(s)	Technology	Production cost
Cabacungan et al., 2009	Car air-conditioning unit as a dehumidifier	$0.20 per liter
Yildirim and Solmuş, 2014	Humidification–dehumidification (HDH)	$0.10 per liter
Milani D. et al., 2014	Hybrid thermoelectric-absorption atmospheric water generator (HTE-AWG)	$0.20 and $0.40 per liter
Alahmer et al., 2017	Refrigeration with high-quality potable water	$0.20 and $0.40 per liter
Jradi, 2011	Integrated Thermoelectric-Photovoltaic system of humid air	$ 0.082 per liter
Munoz-garcía et al., 2013	Using thermoelectric devices for water production	$ 0.40 per liter
Suryaningsih and Nurhilal, 2016	Optimized using Computational Fluid Dynamics (CFD)	$0.20 per liter
Kabeel et al., 2016	Numerical study of a straightforward thermoelectrical device	$0.15 per liter
Joshi et al., 2017	A thermoelectric device that operated with a cooling surface	$0.68 per liter
Atta, 2017	Thermoelectric cooling device for water production in humid areas,	Around $0.67 per liter
Milani et al., 2014	Dehumidification system that employed solar-assisted desiccant for water generation using PV-T	around $0.20 per liter

Conclusion

Solar energy utilization for water production emerges as a sustainable and eco-friendly solution, particularly beneficial in remote or off-grid areas with limited access to electricity. The on-site installation of solar panels eliminates the need for a grid connection, making solar energy cost-effective as technology costs decrease. Moreover, solar thermal desalination systems have the added advantage of generating valuable heat during the desalination process. Space could be efficiently utilized for space or process heating or converted into electricity through thermoelectric generators.

Integrating thermoelectric generators into AWG systems further enhances water production capabilities, even in regions with low humidity or high temperatures, such as arid or desert areas. Considering the production cost data, three highly efficient technologies stand out for water production. Humidification-Dehumidification (HDH) method: $0.10 per liter, optimized Solar-Driven Thermoelectric Dehumidification system: $0.10 per liter, and Thermoelectric-Photovoltaic system of humid air: $0.082 per liter. Three thermoelectric-based techniques stand out for their high-water production capacities. A solar-driven thermoelectric system that has a capacity of 10 liters per day, suitable for generating significant amounts of water in the humid Lebanese climate's solar-driven. The prototype of AWG system based on a thermoelectric cooler (TEC), which achieves an impressive capacity of 14.6 liters per day, providing an efficient solution for water generation.

In conclusion, these advanced thermoelectric technologies, combined with solar energy and other innovative techniques, hold tremendous potential to alleviate water scarcity, deliver sustainable water solutions, and ensure a brighter and water-secure future for communities around the world. As research and development in this field continue to progress, the efficiency and effectiveness of these technologies are bound to improve, making them increasingly valuable in addressing global water challenges.

References

Alahmer, A., Al-Dabbas, M., Alsaqoor S., and Al-Sarayreh, A. (2018). Utilizing Solar Energy for Extracting Freshwater from Atmospheric Air. *Applied Solar Energy*, 54(2): 110-118. https://doi.org/10.3103/S0003701X18020044.

Al-Karaghouli, A., and Kazmerski, L. L. (2013). Energy Consumption and Water Production Cost of Conventional And Renewable-Energy-Powered Desalination

Processes. *Renewable and Sustainable Energy Reviews*, 24: 343-356. https://doi.org/10.1016/j.rser.2012.12.064.

Almusaied, Z., and Asiabanpour, B. (2017). Atmospheric Water Generation: Technologies and Influential Factors. In *proceedings of 67th Annual Conference and Expo of the Institute of Industrial Engineers*: 1448-1453. https://doi.org/10.1007/978-3-030-29996-5_18.

Atta, R. M. (2017). Solar Water Condensation Using Thermoelectric Coolers. International *Journal of Advanced Engineering Research and Development*, 4(09). https://doi.org/10.21090/ijaerd.96616.

Aye, L., George, B., and Wu, D. (2013). Solar Chilled Drinking Water Sourced From Thin Air: Modelling and Simulation of a Solar-Powered Atmospheric Water Generator. In *Proceedings of International Congress on Modelling and Simulation, MODSIM* 2013: 2360-2365. https://doi.org/10.36334/modsim.2013.l2.aye.

Cabacungan, P. M., Cheng, J. C., Landoy, D. L. C., Silang, A. C., Buenafe, R. A., Tangonan, G. L., Libatique, N. J. C., and Calasanz, T. H. (2009). Solar-Powered Atmospheric Water Generation and Purification System. In *Proceedings of 4th International Conference on Humanoid, Nanotechnology, Information Technology, Communication and Control, Environment, and Management*: 138909101.

Chaudhry F. N. and Malik M. F (2017). Factors Affecting Water Pollution: A Review. *Journal of Ecosystem and Ecography*, 7(1): 1-3. https://doi.org/10.4172/2157-7625.1000225.

Dave, T., Ahuja, V., and Krishnan, S. (2021). Economic Analysis and Experimental Investigation of a Direct Absorption Solar Humidification-Dehumidification System for Decentralized Water Production. *Sustainable Energy Technologies and Assessments*, 46: 101306. https://doi.org/10.1016/j.seta.2021.101306.

Dia Milani, A. V., Abbas, A., Chiesa, M., and Al Bakri, D. (2011). Evaluation of Using Thermoelectric Coolers in A Dehumidification System. *Chemical Engineering Science*, 66(12): 2491-2501. https://doi.org/10.1016/J.CES.2011.02.018.

Elattar, H. F., Fouda, A., and Nada, S. A. (2016). Performance Investigation of a Novel Solar Hybrid Air Conditioning and Humidification-Dehumidification Water Desalination System. *Desalination*, 382: 28-42. https://doi.org/10.1016/j.desal.2015.12.023.

Elgendi, M., Kabeel, A. E., and Essa, F. A. (2023). Improving the Solar Still Productivity Using Thermoelectric Materials: A Review. *Alexandria Engineering Journal*, 65: 963-982. https://doi.org/10.1016/j.aej.2022.10.011.

El-Ghonemy, A. M. K. (2012). Fresh Water Production From/By Atmospheric Air for Arid Regions, Using Solar Energy: Review. *Renewable and Sustainable Energy Reviews*, 16(8), 6384-6422. https://doi.org/10.1016/j.rser.2012.06.029.

Esfahani, J. A., Rahbar, N., and Lavvaf, M. (2011). Utilization of Thermoelectric Cooling in a Portable Active Solar Still - An Experimental Study On Winter Days. *Desalination*, 269(1–3): 198-205. https://doi.org/10.1016/j.desal.2010.10.062.

Esfe, H. M., Esfandeh, S., and Toghraie, D. (2021). Numerical Simulation of Water Production from Humid Air for Khuzestan Province: Investigation of the Peltier Effect (Thermoelectric Cooling System) on Water Production Rate. *Case Studies in Thermal Engineering*, 28. https://doi.org/10.1016/j.csite.2021.101473.

Hosseini, S. E., Andwari, A. M., Wahid, M.A., and Bagheri, G. (2013). A Review on Green Energy Potentials in Iran. *Renewable and Sustainable Energy Reviews*, 27: 533-545. https://doi.org/10.1016/j.rser.2013.07.015.

Jabari, F., Nazari-heris, M., Mohammadi-ivatloo, B., Asadi, S., and Abapour, M. (2020). A Solar Dish Stirling Engine Combined Humidification-Dehumidification Desalination Cycle for Cleaner Production of Cool, Pure Water, and Power in Hot And Humid Regions. *Sustainable Energy Technologies and Assessments*, 37:100642. https://doi.org/10.1016/j.seta.2020.100642.

Joshi, V. P., Joshi, V. S., Kothari, H. A., Mahajan, M. D., Chaudhari, M. B., and Sant, K. D. (2017). Experimental Investigations on a Portable Fresh Water Generator Using a Thermoelectric Cooler. *Energy Procedia*, 109: 161-166. https://doi.org/10.1016/j.egypro.2017.03.085.

Jradi, M., Ghaddar, N., and Ghali, K. (2011). Experimental and Theoretical Study of an Integrated Thermoelectric-Photovoltaic System for Air Dehumidification and Fresh Water Production. *International Journal of Energy Research*, 36(9): 963-974 https://doi.org/10.1002/er.1848.

Jradi M., Ghaddar, N., Ghali k. (2012). Optimized Operation of a Solar-Driven Thermoelectric Dehumidification. *Journal of Energy and Power Engineering*, 6: 878-891.

Jury, WA, and Vaux, H. J. (2007). The Emerging Global Water Crisis: Managing Scarcity and Conflict Between Water Users. *Advances in Agronomy*, 95(7): 1-76. https://doi.org/10.1016/S0065-2113(07)95001-4.

Kabeel, A. E., Omara, Z. M., and Younes, M. M. (2015). Techniques Used to Improve the Performance of The Stepped Solar Still - A Review. *Renewable and Sustainable Energy Reviews,* 46: 178-188. https://doi.org/10.1016/j.rser.2015.02.053.

Kabeel, A. E., Abdulaziz, M., and El-Said, E. M. S. (2016). Solar-Based Atmospheric Water Generator Utilization of Fresh Water Recovery: A Numerical Study. *International Journal of Ambient Energy* 37(1): 68-75. https://doi.org/10.1080/01430750.2014.882864.

Khalil, B., Adamowski, J., Shabbir, A., Jang, C., Rojas, M., Reilly, K., and Ozga-Zielinski, B. (2016). A Review: Dew Water Collection From Radiative Passive Collectors to Recent Developments of Active Collectors. *Sustainable Water Resources Management*, 2(1): 71-86. https://doi.org/10.1007/s40899-015-0038-z.

Kode, V. R., Stuckenberg, D. J., Went, E. K., Erickson, O. M., and Plumer, E. (2022). Techno-Economic Analysis of Atmospheric Water Generation by Hybrid Nanofluids to Mitigate Global Water Scarcity. *Liquids*, 2(3): 183-195. https://doi.org/10.3390/liquids2030012.

Liu, Y., Shi, Z., Wang, G., Yan, Y. Y., and Zhang, Y. (2021). Experimental Investigation for a Novel Prototype of a Thermoelectric Power Generator with Heat Pipes. *Frontiers in Energy Research*, 9: 1-13. https://doi.org/10.3389/fenrg.2021.744366.

Mekhilef, S., Saidur, R., and Safari, A. (2011). A Review on Solar Energy Use in Industries. *Renewable and Sustainable Energy Reviews*, 15(4): 1777-1790. https://doi.org/10.1016/j.rser.2010.12.018.

Micari, M., Duan, X., and Agrawal, K. V. (2023). Atmospheric Water Harvesting in Semi-Arid Regions by Membranes: A Techno-Economic Assessment. *Journal of Membrane Science*, 672: 121437. https://doi.org/10.3390/liquids2030012.

Milani, D., Qadir, A., Vassallo, A., Chiesa, M., and Abbas, A. (2014). Experimentally Validated Model for Atmospheric Water Generation Using A Solar-Assisted Desiccant Dehumidification System. *Energy and Buildings*, 77: 236-246. https://doi.org/10.1016/j.enbuild.2014.03.041.

Mulchandani, A., and Westerhoff, P. (2020). Geospatial Climatic Factors Influence Water Production of Solar Desiccant Driven Atmospheric Water Capture Devices. *Environmental Science and Technology*, 54(13): 8310-8322. https://doi.org/10.1021/acs.est.0c00534.

Muñoz-García, M. A., Moreda, G. P., Raga-Arroyo, M. P., and Marín-González, O. (2013). Water Harvesting for Young Trees Using Peltier Modules Powered by Photovoltaic Solar Energy. *Computers and Electronics in Agriculture*, 93: 60-67. https://doi.org/10.1016/j.compag.2013.01.014.

Panchal, H., Sadasivuni, K. K., Prajapati, C., Khalid, M., Essa, F. A., Shanmugan, S., Pandya, N., Suresh, M., Israr, M., Dharaskar, S., and Khechekhouche, A. (2020). Productivity Enhancement of Solar Still with Thermoelectric Modules from Groundwater to Produce Potable Water: A Review. *Groundwater for Sustainable Development*, 11: 100429. https://doi.org/10.1016/j.gsd.2020.100429.

Salehi, A. A., Ghannadi-Maragheh, M., Torab-Mostaedi, M., Torkaman, R., and Asadollahzadeh, M. (2020). A Review on the Water-Energy Nexus For Drinking Water Production from Humid Air. *Renewable and Sustainable Energy Reviews*, 120: 109627. https://doi.org/10.1016/j.rser.2019.109627.

Sandford, B., and Adeel, Z. (2012). *The Global Water Crisis: Addressing an Urgent Security Issue*. Hamilton, Canada: UNU-INWEH. https://doi.org/10.13140/2.1.2593.8721.

Sleiti, A. K., Al-Khawaja, H., and Al-Ali, M. (2021). Harvesting Water From Air Using Adsorption Material - Prototype and Experimental Results. *Separation and Purification Technology*, 257: 117921. https://doi.org/10.1016/j.seppur.2020.117921.

Smith, V. H. (2003). Eutrophication of Freshwater and Coastal Marine Ecosystems: A Global Problem. *Environmental Science and Pollution Research*, 10 (2): 126-139. https://doi.org/10.1065/espr2002.12.142.

Sun, J., Wu, T., Wu, H., Li, W., Li, L., Liu, S., Wang, J., Malfait, W. J., Zhao, S. (2023). Aerogel-Based Solar-Powered Water Production from Atmosphere and Ocean: A Review. *Materials Science and Engineering*, 154. https://doi.org/10.1016/j.mser.2023.100735.

Sun, W., Liu, W-D., Li, L., Wang, D-Z., Yin, L-C., Li, M., Shi, X-L., Liu, Q. and Chen, Z-G. (2023). Performance Optimization of a Dual-Thermoelectric-Liquid Hybrid System for Central Processing Unit Cooling. *Energy Conversion and Management*, 290: 117222. https://doi.org/10.1016/j.enconman.2023.117222.

Suryaningsih, S., and Nurhilal, O. (2016). Optimal Design of an Atmospheric Water Generator (AWG) Based on Thermo-Electric Cooler (TEC) for Drought In Rural Area. *AIP Conference Proceedings*, 1712. https://doi.org/10.1063/1.4941874.

Tu, Y., Wang, R., Zhang, Y., and Wang, J. (2018). Progress and Expectation of Atmospheric Water Harvesting. *Joule*, 2(8): 1452-1475. https://doi.org/10.1016/j.joule.2018.07.015.

Vián, J. G., Astrain, D., and Domínguez, M. (2002). Numerical Modeling and a Design of a Thermoelectric Dehumidifier. *Applied Thermal Engineering*, 22(4): 407-422. https://doi.org/10.1016/S1359-4311(01)00102-8.

Wehbi, Z., Taher, R., Faraj, J., Castelain, C., and Khaled, M. (2023). Hybrid Thermoelectric Generators-Renewable Energy Systems: A Short Review on Recent Developments. *Energy Reports*, 8: 1361-370. https://doi.org/10.1016/j.egyr.2022.08.068.

Yildirim, C., and Solmuş, I. (2014). A Parametric Study on A Humidification-Dehumidification (HDH) Desalination Unit Powered by Solar Air and Water Heaters. *Energy Conversion and Management,* 86: 568-575. https://doi.org/10.1016/j.enconman.2014.06.016.

Chapter 10

Contribution of the Functional Land Usage Organization on the Sustainability of Residential Environment

Fatima L. S. Al-Dairy[1,*]
and Haider J. E. Al-Saaidy[1]
[1]Department of Architectural Engineering, University of Technology- Iraq, Baghdad, Iraq

Abstract

Functional organization of land usage has a wide impact on the patterns of residential environments. The chapter explores how the organization affected most residential environments' final pattern. It addresses the lack of functional organization of land usage due to the association of organization with the three constituent aspects: conceptual, performance, and morphological. Lack of identification has been achieved by defining the concept of the residential environment and its components represented by the conceptual aspect that includes time, place, and activity. The performance aspect includes human needs and how to provide acceptable standards of life for the residents to achieve the possibility of living by controlling the third morphological aspect. The outcome from the interpretation of previous studies declares that the formation of the urban body is directly affected by the functional organization of urban land uses associated with the change in time. Hence, the decline in the level of functional organization leads to a decline in the level of formation and, thus, a decline in performance, leading to positive or negative effects on the residential environment. The importance of the topic lies in the fact that most of the studies dealt with

[*] Corresponding Author's Email: fatimasafi306@gmail.com.

In: Advancements in Sustainability Systems
Editors: Hussain H. Al-Kayiem, Hasanain A. Abdul Wahhab and Raed A, Jessam
ISBN: 979-8-89113-571-0

the definition of order and organization. Still, they did not differentiate between the general system and its subsystems. In addition, most of the studies did not deal with the factors influencing the system and organization, making it an effective and influential element in changing the apparent aspect of the residential environment. Reviewing and analyzing the literature declares that there is a knowledge gap about the effect of organizing in changing the patterns of residential environments.

Keywords: organization, change, function, functional land use, residential environments

1. Introduction

The residential environment is a spatial phenomenon with a complex system of activities linked between them to specific diagnostic and measurable relationships that reach the accuracy of individuals' relations with their preaching and reach the level of inclusiveness to the level of mutual influence between distant areas. Therefore, the study and analysis of the functional system should be based on a set of criteria and values consisting of the interaction of political, economic, social, and natural conditions. In the past, the system was one of the main guiding principles in forming cities. These principles have changed due to weak laws and regulations and a lack of oversight affecting the housing environment. After many definitions and reviewing studies such as (Arnheim, 1977), there has been a knowledge gap about functional regulation's impact on changing housing environment patterns. First, by establishing a clear and specific definition of the concept of change and residential environment and its components, the main objective was to explore the machinery by which functional organization influenced the ultimate pattern of most residential environments. The study found that the changing functional system was due to changing human needs associated with changing times, which in turn reflected the visible aspect of the residential environment of the structural aspect.

1.1. The Change

The term change is called for every movement and interaction. It includes the transition from one state to another, meaning the opposite of the relative stability of objects' qualities, construction, or laws of their existence. Rosenthal, 1981 commented on the "change" that the term change can be called "propulsive change." According to some researchers in the science of civilization, change is an all-encompassing transformation process that may address the nature of the same. At the same time, some sociologists consider that change is based primarily on sudden movement and does not depend on slow time and movement. As mentioned by Hass, 1999, the Arabic scientist Ibn Khaldoun believes that there is a distinction between rapid change and slow change as he referred to it in his introduction: "The historian has to pay attention to the constant evolution and change that is taking place in everything in existence and, therefore, as the present is harsh about the past, he has to pay attention to the disagreement between them, and to whom because the lack of attention to this falls fundamentally wrong (Peruti, 1992). Through the preceding, the change is the transition from one state to another, whether that change is partly, totally, rapidly, or slowly, and if the objects or the laws or attributes all come to mean that it is a process of transformation, change, displacement, or replacement.

1.2. The Residential Environment

In Arabic, the word (environment) refers to the name of the act, i.e., the hostel or residence, and to any repairing and preparation thereof [1](Bekim, 2010), as well as the place where the appropriate factors are available for the livelihood of a living organism or group of living organisms and include the social and natural environment. Alternatively, it is home or a natural, social, or political environment[2]. The first to use the terminological meaning of the environment was the German scientist Haeckelin, 1866 as mentioned by Domokos, 2011, who arrived at its meaning using the term "Ecology," a merger of two Greek words, the first "Oikos," which means dwelling and the second "Logos," meaning science, and defined it as the science that studies the relationship of organisms with the medium in which they live. The environment in the humanities includes the built or created environment, such as the housing environment. The working environment is the total composition of the human community between the natural environment and the built environment, what

[1] http://arabiclexicon.hawramani.com/ibn-manzur-lisan-al-arab.
[2] https://www.almaany.com.

is known as the human environment, their impact on each other, and their compatibility (Algeria, 2015).

The environment in sociology is characterized by the medium or framework in which man lives and is influenced and influenced by it. In geography, the environment is defined as all the factors surrounding organisms that are controlled by social and economic factors and affect their upbringing, evolution, and various manifestations of their lives, as defined by the Professor of Biological Sciences, Ricardos, as "the sum of the surrounding natural factors that affect all organisms and are an interconnected ecological unit."[3] Rapoport refers to a series of relationships between physical and human elements, which are at two levels, between other elements and elements, and between human and human, on the one hand, and between human and human, on the other. The environment is concerned with housing, its tools, facilities, and related services as a place of family residence and activity (Omar, 2011). They are also described as a set of systems that are intertwined with each other to the degree of complexity that affects and determines human survival in this world. The environment is the place where a range of living and non-living ingredients are interrelated with each other. These relationships are complex and interconnected to achieve balance and stability for the components (Bekim, 2010). These relationships are shown in Figure 1.

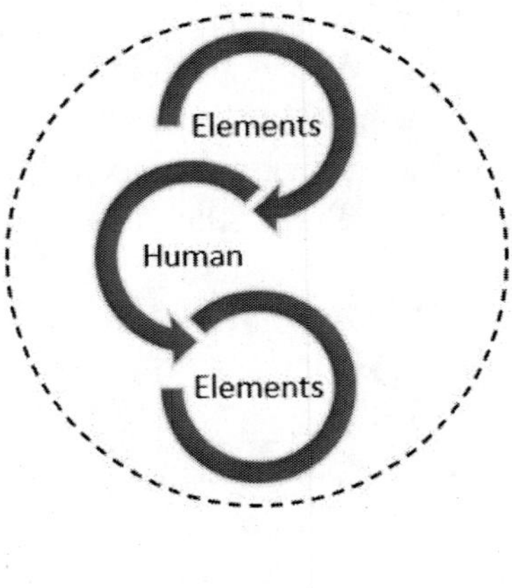

Source: Author.

Figure 1. A series of relationships between man and physical elements.

[3] http://www.ukessays.com.

1.2.1. Components of the Environment

The environment generally consists of two main components: the non-living component of the physical environment (natural, constructive) and the living component of the social environment, as shown in Figure 2.

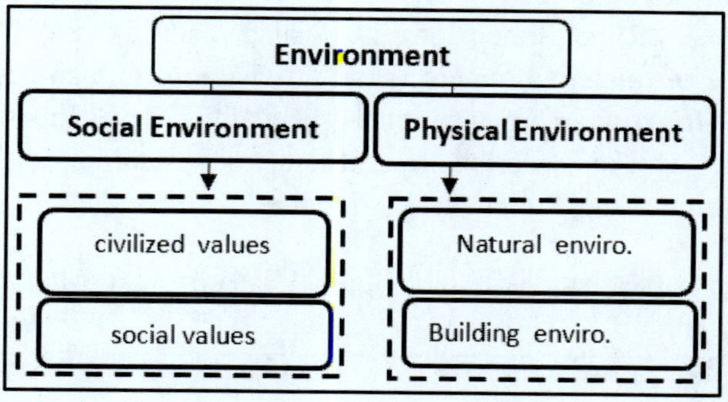

Source: Author.

Figure 2. Components of the environment.

1.2.2. The Concept of the Residential Environment

The concept of the residential environment is addressed in several respects. Dictionary of Science, defined as an area characterized by the existence of housing complexes. It is the medium with certain environmental characteristics, and its inhabitants can choose within cultural determinants associated with their lifestyle. This choice reflects the desire to achieve cultural ideals, values, and perceptions. It is a formal expression of social and political systems. This aspect is important for the residents to improve health, provide income, and security (Ghuli, 2004; Abbas and Salal, 2017).

The place where man interacts with others, where all human characteristics and qualities are associated, and is the ground for creating interaction and social communication. It is also a structure rather than a random assembly of objects, shaped by the identification of classifications that represent building blocks by finding a place for everything in the whole system as well as through the relationships between the human element of the people in that structure and the material element that is all that surrounds these people. The characteristics of the relationship between the two elements and their nature determine what the environment is and define its identity and basic features, which appear in spatial form by expressing them in types of regulations, such as spatial, communication, temporal, and social organization and the organization of meanings. Table 1 defines the elements of the residential environment.

The theoretical definition of the residential environment is the middle or place where human livelihood affects and is influenced by its activities within different periods, highlights the importance of the definition of place and time, which is one of the dimensions associated with the environment. The performance of this environment is the provision of human needs achieved through the structural or organizational aspect of the uses within the residential environment, which is described as a structure or system formed through a set of elements and relationships that is shown in Figure 3.

Table 1. The definitions of the residential environment

Defining the residential environment	Description
• An area mainly characterized by houses and apartment complexes. • The medium possesses certain environmental characteristics, and its residents can choose within cultural determinants related to their way of life, reflecting the desire to achieve ideals, values, and perceptions. • Formal expressions of social and political systems are just as important as what the inhabitant does to improve health, provide income, and generate security. • It is the place where a person interacts with others and creates interaction and social. • It is a structure, not a random grouping of things, expressed in various types of organization, such as spatial, communicative, temporal, and social organization, and the organization of meanings.	• Take up the physical description of the environment. • Dealing with the relationship between the environment and man and how one affects the other. • Represents a form resulting from the reflection of the social behavior of the inhabitants. • Times It was dealt with as the ground on which a system of relationships is based. • It was dealt with as an integrated structure consisting of a group of systems represented by the physical and social systems and the relationship between them. In addition to that, each of these systems includes within it a set of secondary organizations through their interdependence that form the integrated pattern within the residential environment.

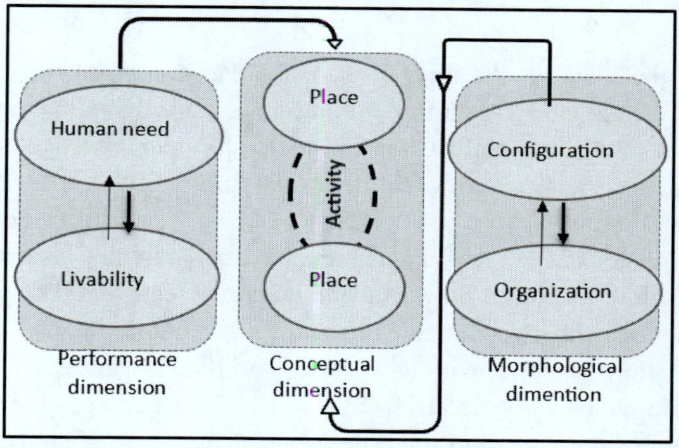

Source: Author.

Figure 3. Residential dimensions and their relationship to each other.

2. Dimensions Related to the Residential Environment

The residential environment is associated with three aspects. One aspect refers to it as a concept that relates to its main components (human, place, and activity). The second aspect relates to the purpose of the residential environment, which is to meet human needs and enable human life. The third aspect is related to how these needs are met through the control of the formation and organization of functions and their relationships to reach the desired performance.

2.1. Conceptual or Intellectual Dimension

The time dimension is essential for understanding life within different spaces. The time a person spends in a street is a major factor for measuring vitality, which is directly proportional to the individual's period of stay within the space. When the individual's period of stay within the space is increased, it is called (long-term) activities that are activated by controlling the material elements or the ground interfaces in a way that encourages pedestrians to stop or walk slowly (Ghuli, 2004). This means that the three elements (time, place, activity) are interrelated. If one of them is changed, the rest will be changed as a result. The meaning of each of these elements will be discussed.

2.1.1. Place

It means the place of the thing's existence and a concept associated with (ability), which means stability in place and enhancing the individual's practical or intellectual ability. It is considered for man to be like the Earth on which he stands and the body that he owns (Hillier, 2007). It includes many meanings, such as space, position, place, being, and space. Hillier, 2007 explained that space is the place where all human activities take place and is considered a reflection of the social and cultural events that take place in the city. It requires understanding the impact of space on people, how they use it, and how they interact with each other. Although space and time are inseparable elements, space is fixed and can be perceived directly by the senses, while time is moving and is perceived by humans indirectly. It also has three dimensions, while time has only one dimension. Space expresses the distribution of existing things automatically, while time expresses the succession of the existence of phenomena, where one replaces the other. Time does not return, meaning that every material process develops in one direction from the past to the future (Mohmoud, 2016).

2.1.2. Time

It is the unlimited continuous progression of existence and events in the past, present, and future. The word (time) means few and many (Bourne, 2007). In its physical meaning, time is the measure of movement, and this means that time is related to movement in that it is the preceding and the subsequent. It is also considered a fourth dimension measured with the three dimensions and is influential in the visual perception of the place (Oxford Dictionary, 2016). Safdie, 1997 sees time through the idea of the event, and had it not been for the change of the event. There would be no sense of time, so the change of events is in the way that consciousness can notice the movement. To understand it, something must happen to it that makes it different from what it was. Time is the framework through which we realize change and change when it deals with material things and expresses its transfer in space in the sense that every process takes place within a framework of time and space (Al-Dulaimi, 2002). Figure 4 demonstrates the relationship between the organization and the time.

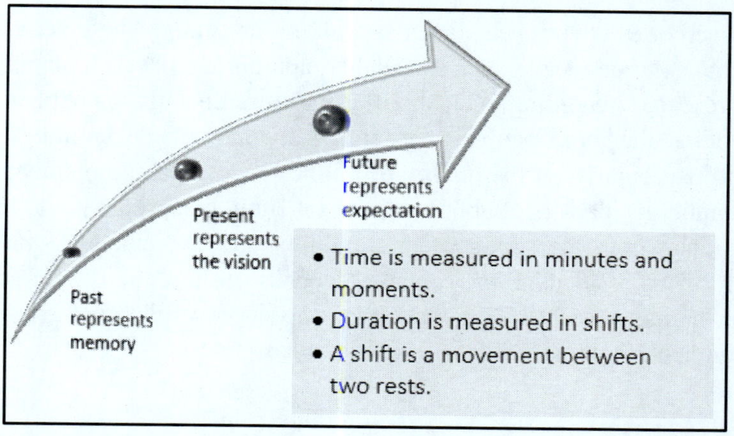

Figure 4. The relationship of organization with the change of time.

From the preceding, the environment organizes time through communication between individuals in the place. Hence, the definition of time is through defining the change in time, and if it indicates the change of that time, space, time, and event are considered inseparable in defining the residential environment. Table 2 summarizes the concepts of time and place.

Table 2. Summary of time and place concepts

Concept	connotations, features
Place	Earth, space, house, body, space, position, position, constancy, empowerment, form of external experience, entity, void, body, universe (length, width, height) A stationary that can be perceived directly by the senses, a fixed abstract void of a nature formed and transformed with time, the form of external experience
Time	Time is little and a lot, period, era, continuous progress, form of inner experience. Moving, aware indirectly, with one dimension, renewed and known, an immaterial continuum, the succession of the existence of phenomena (past, present, future), intuitively linked to the event, the framework through which we perceive change.

2.1.3. Activity

Activity is influenced by functional distribution closely linked to physical factors, and activities can be summarized as (necessary-optional-social). The necessary activities include the day-to-day tasks of the region's inhabitants, such as going to school or work, occurring throughout the year. They are practically independent and an option for residents, often related to walking. Optional activities occur under certain circumstances with people's participation if there is a desire, time, and place to do so. At the same time, social activities cover events such as walking or sitting. These activities are linked to external factors such as weather, appropriate physical planning, and user presence. According to Gehl, 2011, the more time, the more people meet each other, the more people there are. Activities outside buildings, which depend particularly on the quality of public spaces, are complementary and are implicitly part of public activities. They disappear under adverse conditions while flourishing when conditions are favorable. The successful design of the residential environment depends on how the design performs against human needs, which realizes the concept of livability in the residential environment. So those needs will be addressed.

2.2. Performance Dimensions

The concept of performance includes an attempt to determine how the residential environment should perform or achieve the desired result without resorting to describing how the solution or result is supposed to be, and it is considered an assessment of the suitability of the product for the application. Performative rush is a measure of the amount of gratification and satisfaction that a place provides to its occupants and is represented by the following concepts.

2.2.1. Human Needs

An urban design with effective performance meets people's needs by meeting their self-needs consistent with other selves within the urban environment. In addition to the means, they use to move and reach the areas they want to reach. In modernist planning, the single-function residential areas resulting from zoning lacked the necessary dynamism, mobility, and diversity of urban life. This led to traffic congestion, increased per capita trips, and increased population in the commercial cities' centers and other similar individual functional areas at some hours of the day and vacating them at other times. The problem with the modern idea is that how people's needs are manifested could be implemented globally and did not realize the full range of general human needs and the individual differences among people. Maslow and Lewis, 1987 assumed that needs are organized in a gradient or escalating system in terms of priority or intensity of impact prepotency. When the most priority needs are achieved, the following needs also demand satisfaction, as shown in Figures 5 and 6. Then, a higher degree on the ladder of motives is climbed, and so on, until reaching the top (Oxford Dictionary, 2016).

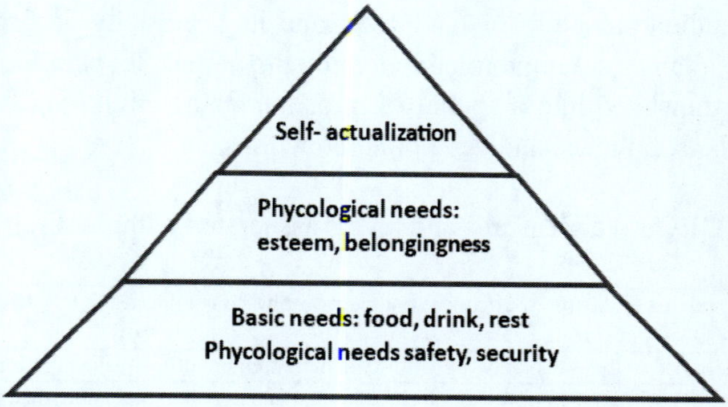

Source: Authors.

Figure 5. Human needs and motives in Maslow's hierarchy.

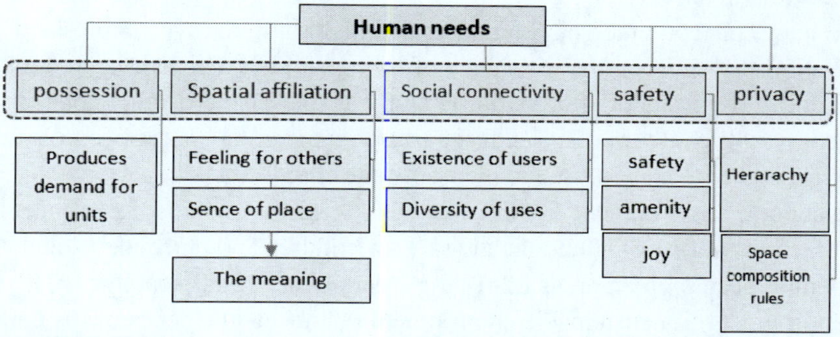

Source: Authors.

Figure 6. Human needs and how to achieve them within the residential environment.

Most of the human needs related to the urban dimension are related to land use and levels that could be summarized in four urban measurements, namely Micro, Macro, Local, and Global levels, as shown in Table 3. The first dimension, Micro, is related to the miniature urban environment with the boundaries of the street frontage or a specific residential area. In contrast, Macro is related to the urban network within the measurement of the residential locality. While (Local) covers a wider area of the urban environment, it may combine more than one residential locality within one urban network. At the same time (Global) includes a relatively large urban area such as a city. These measurements or levels are relative and related to the specificity of the research and the selected study sample.

The studied signature of land uses, and its specificity according to the urban level plays a strategic role in meeting the highest level of human needs. These are translated into a spatial form or land use activity and reflected in land use plans, policies, and development actions.

Table 3. Relative measures and levels associated with research privacy

Micro	It is related to the micro-urban environment within the boundaries of the street frontage or a specific residential area.
Macro	It is linked to the urban network within the measurement of the residential locality.
Local	It covers a wider area of the urban environment and may combine more than one residential locality within one urban network.
Global	It includes a relatively large urban area such as a city.

2.2.2. Livability

A new concept emerged to understand the right to live, expressing the relationship between the efficiency of living and not. It is considered a sub-factor of the issue of sustainability, and it is one of the parallel concepts of quality of life (Bourne, 2007). It focuses on integrating ideas of contentment, dissatisfaction, happiness, and their opposites (Oxford Dictionary, 2016). The residential environment should be like a family, as it is for all people, attractive, worthy of attention, and safe for children and older adults, enabling them to access green areas easily for entertainment and meetings. A livable environment is a sustainable environment that meets the needs of the current population without reducing the ability of the future generation to meet their needs, as projected in Figure 7, (Al-Dulaimi, 2002). The residential environment is not for those who earn money and live outside it. It is also one of its components to having a healthy life and opportunities to move around on foot and by bicycle. By air and by using public transportation or the car when there is no other option. Justice, dignity, ease of access, coexistence, participation, and empowerment are the basic principles of susceptibility to life (Bekim, 2010).

In conclusion, the concept of livability revolves around creating a better environment for people through the people themselves, as it tends towards subjective personal indicators taken from people because they are aware of the goals and know the problems.

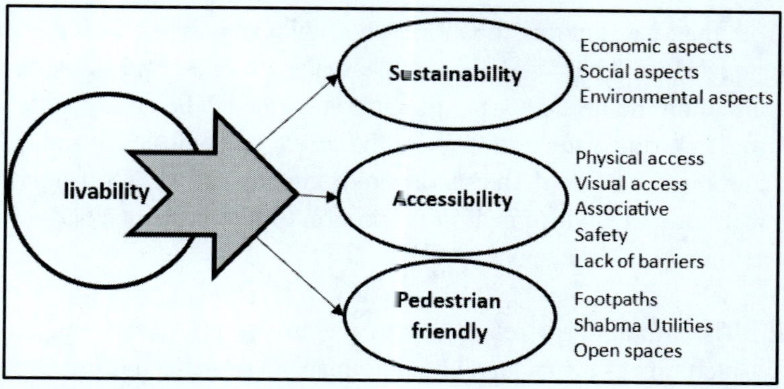

Source: authors.

Figure 7. Basic qualities of livable environments.

From the previous, the residential environment is considered a temporal phenomenon because it possesses the characteristic of transformation and changes over time, according to what is required by the needs of society and spatial, because that transformation affects the place and its characteristics such as clarity, connection, access, etc. It also includes two aspects: material and objective. They are related to human needs, such as feeling comfortable and safe. It is impossible to separate its parts. It requires studying it as an integrated system to understand its formal characteristics and organizational principles resulting from the relationships between the basic organizations that make up the residential environment.

2.3. Morphological Dimension

The morphological dimension can be defined as the science that deals with the study of shape and structure. The concept of urban morphology is the structural fabric of cities, whether small or large. It includes plan, form, function, and the methods by which this fabric is studied, which expresses the forms of cities through historical stages. (Carter, 1982) described it as the study of the general structure of the city, and he linked it to three variables: the plan, the urban fabric, and land use, as these variables interact with each other to produce many varied shapes and models for the city's structure. The urban environment and what gives the city its shape and specificity is the nature of the relationship between the variables (Carter, 1982).

From that, we find that urban morphology is concerned with the study of the organizational characteristics of the space structure and is based on the rules on which the urban space structure is formed. The community plays a major role in formulating it in terms of the extent of its influence and influence on the urban structure of the urban environment and then formulating the urban form of that environment in proportion to that influence and influence. The urban includes three components.

1. The urban form refers to the pattern or physical arrangement or planning of urban land uses, groups, and social activities.
2. The flows and interactions that link uses and activities
3. The spatial or organizational logic or the logic that determines how the different parts fit together (Gehl and Svarre, 2011).

From that, we find that this aspect is divided into two parts: the organizational and formal sides. The organizational aspect will be dealt with first because the organization is the most general, where the formation is considered one of the foundations on which the organization is based (Haas, 1999).

2.3.1. Formation Side

One of the objective of shaping the residential environment is to search for basic planning units. Basic planning units, as many approaches and ideas call for a balance between needs and capabilities, and then many attempts emerged to find an urban unit that fits the human scale and meets the goals and provisions of the relationships between the elements of the formation. The basic unit can be considered as the link between the upper and lower planning levels in the structural composition of the residential environment through what it includes of units of social harmony, which studies have indicated as (the unit associated with a specific population and has integrated and independent services with a minimum of its services, with functional integration between the locations of activities, functions, crossings of movement and communication (components of formation). As an entrance to the formulation of the final formation, the quadrilateral cells appeared.

The residential neighborhood of a social and service center that includes playgrounds, places of worship, the rest of the other public buildings, movement crossings (roads and movement paths) as a functional element and considering the best means of movement within the unit to be on foot, with the necessity of separating motorized traffic from pedestrian movement (on foot). Then, it was suggested that the unit have a network of main roads planned for automatic traffic that surrounds it from the outside and does not penetrate it, provided that the internal movement paths are allocated for local and slow traffic. A distance not exceeding half a mile (800 meters).

It was proposed to separate the residential and commercial areas, if places for the latter be chosen in the corners and allow access to them easily and easily, and designed so that they constitute an independent commercial center. The population density is considered a determinant of the neighborhood, where it was indicated that it is necessary to use low densities of 10 families/acre. From them, it was possible to determine the area of the neighborhood, about 160 acres.

Perhaps the most important components of urban formation can be limited to a group of fixed and recurring functions represented in land uses and their main distributions, namely residential areas, community services, commercial areas, industrial areas, and others. Therefore, those uses, and their characteristics will be addressed to prepare for the research problem.

2.3.2. Land Use

The study of urban land use is of great importance, as it crystallizes through it the functional field of the various activities and events of the city. As a result, its personality and the foundations of its planning are formed. Its special composition is the result of the interaction of three main factors: the human relationship with the surrounding environment, the city's economy, and the type of activities and events that humans practice (Long and Nelson, 2012).

Land use is a broad and relatively complex concept, as it brings together the interactive relationship between human and urban land, which leads to a state of continuous change among them to reach a state of balance, as defined by the International Food and Agriculture Organization (FAO). According to FAO, land use is the function or goal based on which human forces on use the land. Al-Janabi, 1987, one of the scholars in the geography of cities defines it as the human activity that is directly related to the place and defines the activity as the activity practiced on a specific area unit (Mobaraki and Vehbi, 2022). It is also defined as any fixed or periodic type of human intervention to secure material and moral needs, or both, from natural or industrial resources (Maslow and Lewis, 1987). One of the four morphological elements that make up the urban structure includes land use, building structure, plot pattern, and cadastral (street) pattern. It is represented by the changes that occur in the new incoming and current uses that are transferred to other areas. Table 4 summarizes the theorists on land uses.

Table 4. General definitions of land uses and features

Source	Definition of uses	The general characteristic of each definition
Mahmoud, 2016	Represents the interaction between man and urban land over time, The object or purpose for which the Earth was used.	Linking uses with time and considering it as a system of changing relationships.
Carmona, 2003	One of the morphological components of the urban structure.	Consider it part of the urban structure.
Al-Samarrai, 2004	Fixed or periodic type of human intervention for the purpose of satisfying one's physical and moral needs or both natural and industrial resources	Link it to human needs.
Khalifa, 2004	The visual dimension of land exploitation	linking it to the urban landscape.
Al Janabi, 1987	Human activity associated with the place	linking it to the physical structure.

Source: Prepared by Author.

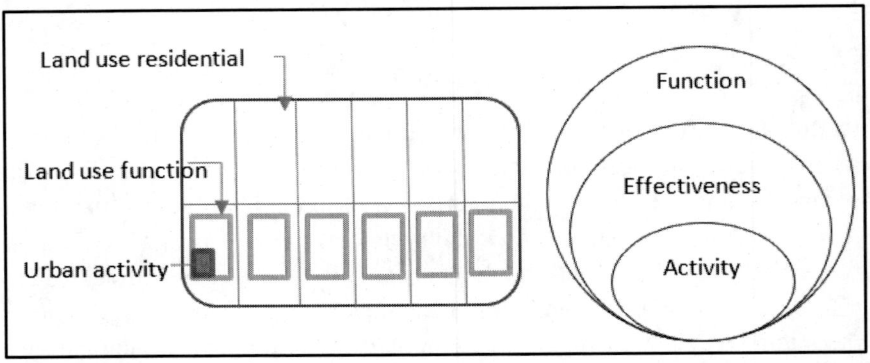

Source: Author.

Figure 8. The functional hierarchy of uses.

Among the definitions, it was found that land uses are related to (place, people, and time) and change according to the changing human needs that change over time.

The concept of land use is always related to functional activities (Function), as displayed in Figure 8, which expresses the use of the space in which human activity is practiced (Activities), as the land uses include within a set of functions according to the study of Ling and Shu-Kui, 2010. The study stated that the land uses consist of five functions, including:

- residence functions or housing, service functions, including commercial, industrial, health, and entertainment functions,
- infrastructure functions or environmental protection functions such as wastewater treatment, energy, and land provision,
- external spaces functions that meet the population's spiritual needs of sense, touch, and vision, and functions security associated with urban management.

However, it should be noted that there are several concepts related to the job, such as Function, Effectuality, and Activities. The concepts of land use related to the job are differentiated in Table 5. The comparative analysis in Table 5, it could be concluded that the relationships that link them are considered a system that cannot be dealt with in part. The components of the city represented here (the functional system) cannot be dealt with in isolation, each function from the other, because of the mutual influence of these functions or uses among them. Therefore, the planning of these uses takes place on two levels.

Table 5. The difference between function, potency, efficacy, and activity.

Concepts	Meanings
Function	It is the main purpose for which the building was designed, for example, service functions: it includes commercial, industrial, recreational, and so on.
Effectiveness	The production is determined or desired. It gives the meaning of efficiency when it expresses the degree of success of something in achieving the desired result, which is often related to a percentage of the degree of success of that thing, and it differs from effectiveness.
Performance	which is defined as the correct performance of that activity. That is, efficient spaces are always active, but active spaces may not be efficient.
Activity	The activity takes place in a place, in a time frame, and often, it is related to the behavior of people inside that place. For example, within the market activity, shopping takes place, meaning that the activity is (a place + a specific performance)

Source: Prepared by Author.

2.3.3. Classification of Land Use

Classification, in general, is the arrangement of shapes or objects into groups and classes according to their common characteristics, and these characteristics determine the place of those things in the classification. Classifying land uses means the description, collection, and planned and organized arrangement of the general and common uses characteristics in a specific area and according to a specific standard. It should be noted that there are no uses placed within a list that represent the uses of the land, but rather, there is a reality, and then it is gradated in importance. Reality determines the uses because it represents the real differences between the components of one land and another.

Among the most important classifications created by some researchers and planners in their study are:

1. Anderson, 1965 discussed Chapin's 1963 concept and how he classified the uses into three functional areas, namely, the housing areas and the various residential uses in addition to other associated services such as parks, play areas, and schools. Work areas include commerce of all kinds, in addition to services, and it is preferable to be close to housing and provide easy access. Entertainment areas, which include cultural and entertainment centers and universities. They should be protected from crowding problems and be close to attractive places. (Chapin, 1963) believes that the uses survey deals with the classification and registration of used spaces, and therefore, he classified them according to the system of urban activities (living, working and leisure).
2. Pacione, 1990 referred to Donald and Arthur, 1961 classification of land use into several categories (residential, commercial, industrial, communications and transportation, and services, which include (health, education, culture, and entertainment) in addition to brown and green fields.
3. Jacobs, 1963, referred to by Lindahl, 1971, classified the uses into primary and secondary. The main uses are represented by residential use, and the rest of the service represents the secondary uses.

2.4. Organizational Side

The organization has been addressed through many concepts and has been defined by various studies as follows:

1. Litterer, 1973 defined the organization as a set of interactions that occur within a specific arrangement. It is also a set of sequential processes intended to arrange new and original social behaviors by adopting the tools that man uses to find integration within the general structure of the society system. Within every organization, there is a

group of systems. Systems, which may bear the characteristic of participating in the formation of more than one organization, include a group of elements linked by certain relationships, and the description of these relationships is that they are mutual interactions and influence between them and that the interactions between these systems represent the form and type of organization to arrange the general structure of the forces of society Order With the intent of controlling social behaviors, that is, unifying the movement of diversity and heterogeneity that exists within the components of that structure, which is represented by three aspects, the social, which is the population, and the cultural, which is knowledge and technology, which are the material means (Lindahl, 1971).

The specificity of the organization that specializes in human activities related to the place, which has certain communication means represented by the points of functional convergence between its elements and components of its elements (systems and system components), which qualify them to be concentrated and gathered spatially. The first step in an organization is to distinguish its most important activities and activities to know its specificities, and then to distinguish its relationship with the individual through communication channels. Because the activities and events related to the individual take a spatial dimension, a physical form is formed for them, which is the urban environment, and the means of communication will be the center of movement and interaction between these activities. This means that, in order to understand the residential environment and its connections, it can be considered an integrated system that arises from the interaction of several systems represented by the physical, social, and technical systems. These systems are linked by relationships with each other, which later form the integrated whole.

2. In his thesis, Arnheim, 1977 linked the concept "system" to the human perception of the urban environment. He pointed out that the system that can be perceived reflects a latent system that may be natural, social, or cognitive. The external system often represents a reflection of the internal system, and any lack of connection between the external and latent systems is what makes the viewer discover that there is a flaw in the entire system. It is also a prerequisite for survival, so it is a strong motive or incentive to produce arrangements regularly through evolution. Disorder is not the absence of order, but rather, it

is represented in the conflict between inconsistent systems. He also believes that the system is a necessary condition for the function of the structure shown in Figure 9.

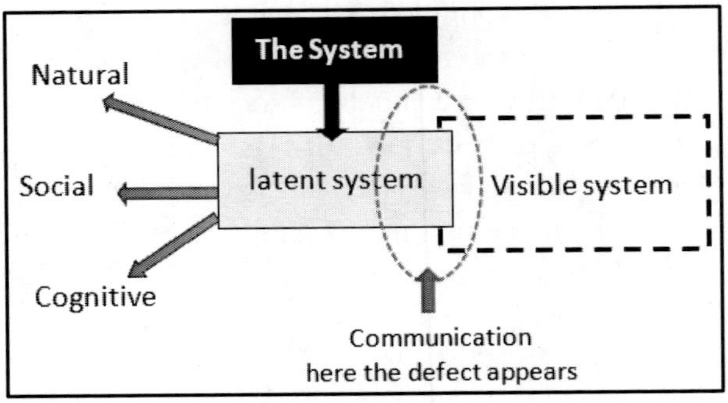

Source: Authors.

Figure 9. System components.

In 1977, Arnheim stated that it should be known that the system is indispensable for producing an organized system. The system is possible at any level of complexity. He believes that the system has levels. The lowest level presents identical housing units, which he calls subdivision, or perhaps known as the distribution of subdivision square plots, which in most cases produce or segregate similar homes that the visitor to the place finds wherever he goes. To avoid this painful situation, it is necessary to limit the trend toward the system by taking advantage of what he called "the theme of the system." He offered three suggestions for correcting the functioning of any system:

- The invalidation of the system is through the non-similarity of functions.
- Everything has independence and the possibility of completion by itself without losing the importance that it will remain part of the larger context.
- Perfection is cold and lifeless.

3. It is noted that the two words "system" and "order," which come with the meaning of "system" in Arabic, are completely different. The word "system" means a set of elements that are internally linked functionally and structurally. These elements appear in the same system within complex relationships, forming a structure or structure

with a unique behavior, as well as including a specific goal. The word "order" comes as an indication of a fixed and continuous relationship between the elements that make up the system. The word "order" arises as an indication of the repetition of basic and fixed relationships or because of the arrangement of adjacent elements. The word "order" seems to create a "system," and perhaps it comes here most often in the sense of the meticulous arrangement of specific parts within a specific system. The word "system" is used when dealing with a structural scale and at the planning level, while the word "order" is used to deal with a smaller scale and at the design level. Hence, the "system" organizes the planning process, while the "order" organizes and generates the design process. Those abbrevaiations have been defined in Table 6. The system is perceived through the concept of centers, while the system can be sensed through the concept of totality (Abo Saada, 2018).

Table 6. Identification of the system, order, Organization, and Communication

Abbreviation	Properties
System in general	• The system's invalidity is through the asymmetry of functions that some of the loss of the system and departure from it can give life to the living environment. • The system can be latent (natural, social, cognitive) and apparent or external. A defect appears when there is a lack of communication between the internal and external systems. • The system is done in the form of step-by-step operations and not in the form of a preliminary plan. • The system is achieved when there is a perfect balance. • The system creates life because of the physical and computational environment. • The system is of two types: structural and geometric. The structural is random, arbitrary, and chaotic, while the geometric depends on lines, levels, and shapes.
Systems	• It is a group of systems with the characteristic of participating in forming more than one organization, and it includes a group of elements linked by certain relationships. These relationships describe that they are mutually interacting and influencing each other. • A group of elements linked internally, functionally, and structurally, and these elements appear in the same system within complex relationships, forming a structure or structure that has a unique behavior, in addition to including a specific goal, used to deal with the structural scale and at the level of planning, and is realized through the concept of centers.

Table 6. (Continued)

Abbreviation	Properties
Order	• Order is a set of interactions that occur in a specific sequence. Or it is a set of sequential operations intended to arrange behaviors by adopting the tools that people use to find integration within the general structure. • It denotes a stable and continuous relationship between the components of the systems. An indication arises of the repetition of simple and fixed basic relationships or as a result of the arrangement of adjacent elements, and it comes in the sense of the elaborate arrangement of certain known parts within a specific system. It is used to deal with a smaller scale and at the level of design through the concept of wholeness. • Order is necessary for making structure, vitality, and arrangement function. • Disorder is not the absence of order but rather the struggle between inconsistent systems.
Organization	• It is an organization formed through the interaction of many constituent elements of subsystems to eventually form the body of the general order.
Communication	• The connections represent the functional meeting points between the elements and their components.

The order is part of the overall system that is formed because of the arrangement of the elements and relationships. Therefore, any defect that affects the order of parts will cause a defect and disturbance to the overall system. Alexander, 1977 who is considered one of those who have a special understanding of the nature of the system and its relationship with the urban environment, believed that the physical urban system is implemented in a procedural manner and not from an initial plan or a preliminary scheme. These operations allow the whole to gradually emerge from local actions, where the system is achieved when there is an ideal balance between the needs of the parts and the needs of the whole.

The study of Alexander, 1977 and Arnheim, 1977 concluded that the system generates life, but they differ in how the system is achieved. Alexander, 1977 points out that the sum of the parts generates the integrated whole, which is one of the stages of the system, which is done in a process or stages. Arnheim, 1977 showed that the system is generated through the compatibility between the latent and manifest systems.

4. Fistola et al., 2020 detected the causes of weakness or entropy within the urban system. It has been interpreted as the result of the interaction of many factors that can alter the system's balance due to its impact on physical, spatial, human, and functional dimensions. The study classified risks into two categories, the first being natural hazards

(such as disasters) and the second including multiple human risks derived from activities that operate simultaneously within the urban system and lead to interactive relationships that are either direct or indirect. They identified immediate risks as affecting the system directly without intermediate stages, while indirect risks could mediate the elements of changing the urban system's functioning. The study also explained that these two types of risks often affect subsystems through interactions within the complex dynamic system. These interactions can affect relationships or their negative effects, which can multiply and intensify, causing the so-called domino effect, which states that incidents result from chronological failure. They stated that competition between elements of regulations leads to the concept of hostility, which is an essential feature of the system's behavior but inevitably leads to potential disruption. When the system falls into crisis, chaos spreads and occurs when differences become conflicts and complementarities into contradictions. The study also showed that entropy production is imperative for developing any complex system (Fistola et al., 2020). This study intends to demonstrate that this production must be contained within the thresholds that limit the system's survival (Fistola, 2011). The study showed that the concept of entropy is useful for understanding and describing the behavior of complex systems. It can be applied to measure the level of organization versus chaos, uniformity versus diversity, useful versus useless, sustainability versus unsustainability, and order versus disruption. "Negative interactions" between components lead to an imbalance that increases entropy, leading the urban system to a structural crisis (Cabral et al., 2013).

5. Abdul-Jabbar and Hassan, 2022 showed that the deep system that produces life directly results from the physical and computational physical structure that occurs in space. He proposed the term "structural system" to completely differ from the "engineering system" that relies on lines, levels, and shapes. The structural system is random, arbitrary, and chaotic, but underneath it is a system, and this hidden system is what creates the living environment. The city has a living structure or structural order that tends to be legible; the living structure emerges slowly, step by step, and as the process proceeds step by step, there is continuous feedback, allowing the system to be directed towards greater wholeness, coherence, and adaptation. Alexander, 1977 considered the city an integrated

building system, and that this system works according to smaller and implicit systems, and it is difficult to predict its future behavior. Furthermore, the ruler by which the nature of the city can be measured in terms of the order, or the absence of order, does not exist, just as the city's entity constantly changes with time, and it is not possible in any way to retrieve its first state. Table 6 summarizes the definitions of the system and order and the linkage elements between them.

Conclusion

It turns out that the system consists of the underlying structure behind a set of perceived relationships. The structure is a system of fixed internal relations that determine the essential features of any entity and form an integrated whole that cannot be reduced to the sum of its elements and a system that governs these elements about how they exist and the laws of their development. From here, we understand that the integrated construction of a thing is the product of the elements and relationships that govern this construction, just as the analysis of any text to understand that structure and hierarchy govern its meaning between the elements and relationships. In this sense, the residential environment can be considered an order system consisting of a group of secondary systems represented by the three systems that make up the residential environment (physical, social, and functional). These systems are based on relationships and are governed by a specific organization. These invisible relationships are represented by time, which affects human needs, which change in turn, leading to a change in the underlying functions or structures, which leads to a change in the final formation, and as a result, the visual appearance of the residential environment changes.

References

Abbas, S. S., & Salal, S. S. (2017). Residents' Satisfaction in the Residential Environment. *The Iraqi Journal of Architecture and Planning*, 16(1): 37-52. https://doi.org/10.36041/iqjap.2017.159701.

Abdul-Jabbar, Z. A.-S., & Hassan, A. B. A. (2022). Spatial Analysis of the Effect of Housing Unit Fragmentation on Infrastructure Services In The City of Baghdad: A Comparative Study Between Al-Salam Neighborhood And Al-Khadra Neighborhood. *Al-Adab Journal*, 3(141): 251-278. https://doi.org/10.31973/aj.v3i141.3724.

Abo Saada, H. J. (2018). *Urban Design Trends- Order and Hope Towards a City*. Academic Library, Cairo, Egypt.

Al-Dulaimi, K. H. A. (2002). *Urban Planning / Foundations and Concepts*. Dar Al-Ilm and Culture for Publishing and Distribution, Gordan. www.academia.edu/39817261.

Alexander, C. (1977). *A Pattern Language: Towns, Buildings, Construction*. Oxford University Press.

Algeria, A. L. (2015). *The Legal System for Environmental Protection and Economic Development*. Dar Houma for printing, publishing and distribution, 23. https://www.eelf.info/tl_files/uploads/EELF%20Books.

Al-Samarrai S. K. (2004). *Spatial Distribution of Land Uses in the Rasasi Al-Irwani Project*. Master's Thesis, College of Education, University of Baghdad, Iraq. http://thesis.mandumah.com/Record/326920.

Anderson, T. R. (1965). Urban Land Use Planning. F. Stuart Chapin, Jr. *American Journal of Sociology*, 71(1). https://doi.org/10.1086/224017.

Arnheim, R. (1977). *The Dynamics of Architectural Form 1977*. ISBN 0520035518, 9780520035515, University of California Press, USA.

Bekim, K. (2010). *Measuring Multifunctionality of Urban Area: Advanced GIS Analysis for Measuring Distance Density Diversity and Time of Urban Services*. Saarbrücken: Lambert Academic Publishing, Germany.

Bourne, L. S. (2007) Understanding Change In Cities: A Personal Research Path. *The Canadian* Geographer, 51: 121-138. https://doi.org/10.1111/j.1541-0064.2007.00170.x.

Cabral, P., Augusto, G., Tewolde, M., and Araya, Y. (2013). Entropy in Urban Systems. *Entropy*, 15: 5223-5236. https://doi.org/10.3390/e15125223.

Chapin, F. S. (1963). *Urban Land Use Planning. Journal of Urban*. University of Illinois Press. USA

Carmona, M., Heath, T., Oc, T., Tiesdell, S. and Carmona, M. (2003). *Public Places - Urban Spaces* (1st ed.). Routledge. London, UK. https://doi.org/10.4324/9780080515427.

Carter, H. (1982). *City and Society: An Outline for Urban Geography*. SAGE Publications, London, England. https://doi.org/10.1177/030913258400800105.

Casey, E. S. (1997). *The Fate of Place. A Philosophical History*. University of California Press. USA. https://doi.org/0.1016/S0962-6298(02)00024-0.

Doxiadis, C. A. (1959). The Rising Tide & The Planner. *Ekistics*, 7(39), 5-10 https://www.jstor.org/stable/43613288.

Fistola, R., Gargiulo, C., and La Rocca, R. (2020). Rethinking Vulnerability in City-Systems: A Methodological Proposal to Assess Urban Entropy. *Environmental impact assessment review*, 85:106464. https://doi.org/10.1016/j.eiar.2020.106464.

Fistola, R. (2011). The unsustainable city. Urban Entropy and Social Capital: The Needing of a New Urban Planning. *Procedia Engineering*, 21: 976-984. https://doi.org/10.1016/j.proeng.2011.11.2102.

Gehl, J., and Svarre, B. (2013). *How to Study Public Life*. Island press center for resource economics. Springer Nature, Singapore. https://doi.org/10.5822/978-1-61091-525-0.

Gehl., J. (2011). *Life between buildings. Using Public Space*. Island Press, Washington, USA.

Ghuli, M. (2004). *Studying and Analyzing the Functional Interaction Between Urban Land Uses*. Ph.D. Thesis, Higher Institute of Urban and Regional Planning for postgraduate studies, Baghdad University, Iraq.

Haas, B. K. (1999). Clarification And Integration of Similar Quality of Life Concepts. *The Journal of Nursing Scholarship*, 31: 215-220. https://doi.org/10.1177/2333721419842671.

Hillier, B. (2007). *Space is the Machine*. Space Syntax, London, UK. https://discovery.ucl.ac.uk/id/eprint/3881/1/SITM.pdf.

Lindahl-Kiessling, K., Alm, G., and Hanna, M.G. (1971). Morphological and Functional Aspects of Immunity. In *Proceedings of the Third International Conference on Lymphatic Tissue and Germinal Centers* held, Uppsala, Sweden, September 1-4, 1970. https://doi.org/10.1007/978-1-4615-9011-8.

Litterer, J. A. (1973). *Analysis of Organizations, 2nd edition*, Wiley, USA. https://www.google.iq/books/edition/The_Analysis_of_Organizations/B7TQzgEACAAJ?hl=ar&bshm=rimc/1.

Long, J. and Nelson, T. N. (2012). Time Geography and Wildlife Home Range Delineation. *The Journal of Wildlife Management* 76: 407-413. http://doi.org/10.2307/41418283.

Ling, L., and Shu-Kui, T. (2010). Functional Analysis of the Land-use System in Urban Residential Areas. In *Proceedings of International Conference on E-Business and E-Government*, Guangzhou, China: 5191-5194. https://doi.org/10.1109/icee.2010.1301.

Mahmoud, W. R. (2016). Time and Space Duality in Interior Design Islamic Architecture Interior Space between Effect and Affected. *Journal of Architecture, Arts, and Humanities*, 1: 384-409. https://doi.org/10.12816/0036530.

Maslow, A., and Lewis, K. J. (1987). Maslow's hierarchy of needs. *Journal of Infrastructure, Planning, and Policy*, 14(17): 987-990. https://doi.org/10.24294/jipd.v6i2.1483.

Mobaraki, A., and Vehbi, B. (2022). A Conceptual Model for Assessing the Relationship between Urban Morphology and Sustainable Urban Form. *Sustainability*, 14: 2884. https://doi.org/10.3390/su14052884.

Oxford Dictionary. (2016). https://global.oup.com/academic/search?q=function&cc=us&lang=en.

Omar, A. (2011). The Concept of Housing and the Impact of Housing. *Baltic Journal of Real Estate Economics and Construction Management*, 4(1): 168-179. https://doi.org/10.1515/bjreecm-2016-0013.

Peruti. F. A. A. (1997). *The architectural development of the house in Baghdad during the twentieth century, with a focus on cultural and social influences*. PhD Thesis, University of Baghdad, Iraq. https://iqdr.iq/search?view=b00a1fb8174b1ada29ab32a9b4b232fa.

Pacione, M. (1990). Urban liveability: a review. *Urban geography*, 11(1): 1-30. https://doi.org/10.2747/0272-3638.11.1.1.

Perry, C. A. (1929). City planning for neighborhood life. *Social Forces*, 8,(1): 98-100. https://doi.org/10.2307/2570059.

Rosenthal, B. (1981). *The Philosophical Encyclopedia*, Dar Al-Tali'ah for Printing and Publishing, Beirut, Lebanon.

Safdie, M. (1997). *The City After the Automobile: An Architect's Vision* (1st ed.). Routledge, London, UK. https://doi.org/10.4324/9780429496752.

About the Editors

Professor Dr. Hussain H. Al-Kayiem
Hilla University College - Iraq
prof.hussain@acaress.org

World's Top 2% Scientists 2022

Professor Hussain completed his BSc in 1973 and MSc in 1981 in Mech. Eng. from the University of Baghdad. After being awarded a PhD from the University of Bradford-UK in 1989, he was appointed as the head of the Mech. Eng. Dept. in the Military College of Engineering – Baghdad for 1990-1997. From 1999 to 2003, he headed the Mech. Eng. Dept. in Al-Mustansiriyah University, Baghdad. He operated as the manager of the Netherlands Engineering Consultants, NEDECO - Iraq branch 2003-2005. Prof Hussain joined Universiti Teknologi PETRONAS (UTP) in Malaysia in 2006 till 2023. He is the founder of the Solar Thermal Advanced Research Center in Malaysia.

As a Thermofluids and Energy Technologies leader, he has published more than 300 papers, 13 chapters, and 8 books. He supervised over 75 PhD and MSc holders. Professor Hussain has been awarded at different national and international levels for his innovative achievements in the energy sector and conferred the 'Eminent Scientist Award' of Wessex Institute – UK.

Associate Professor Dr. Hasanain A. Abdul Wahhab
University of Technology - Iraq
20085@uotechnology.edu.iq

Dr. Hasanain Adnan Abdul Wahhab holds a BSc in Mechanical Engineering, 1996; MSc in mechanical engineering – Combustion, 2000, from the University of Technology; and PhD in Mechanical Engineering - Combustion and Fuel Technology from Universiti Teknologi PETRONAS - Malaysia in 2018. He is a lecturer in the mechanical engineering department at the University of Technology-Baghdad. He has been a Member of Iraqi engineers' union since 1996. Dr. Hasanain authored and co-authored nearly fifty papers published in prestigious international journals and conferences. His research fields focus on renewable energy, automotive, power generation, fuel technology, multi-phase flow, and thermofluids. Currently, his research focus is on combustion operations and magnetization fluids, as well as solar systems.

Associate Professor Dr. Raed A. Jessam
University of Technology - Iraq
50097@uotechnology.edu.iq

Dr. Raed Abbas Jessam received a BSc degree in Mechanical Engineering in 1999 and an MSc degree in Power Systems in 2005 from the University of Technology- Iraq. He is a recipient of a PhD degree in 2018 in Mechanical Engineering, Energy from Universiti Technology PETRONAS- Malaysia. Dr. Raed is an academic staff member at the Electromechanical Engineering Department, University of Technology- Iraq, teaching many undergraduate courses in Mechanical Electromechanical Engineering. He has participated in eight international conferences. He published quality papers in reputable international journals. Along with the academic job, Dr. Raed occupied different positions, such as a member of the exam office for both undergraduate and postgraduate. The areas of research interest of Dr. Raed are energy, renewable energy, thermofluids, thermodynamics, CFD simulation and analysis, and gas turbine technology.

Index

#

A

B

C

D

E

Q

R

S

T

U

V

W